Advanced Concepts for Renewable Energy Supply of Data Centres

RIVER PUBLISHERS SERIES IN RENEWABLE ENERGY

Series Editor

ERIC JOHNSON
Atlantic Consulting
Switzerland

Indexing: All books published in this series are submitted to Thomson Reuters Book Citation Index (BkCI), CrossRef and to Google Scholar.

The "River Publishers Series in Renewable Energy" is a series of comprehensive academic and professional books which focus on theory and applications in renewable energy and sustainable energy solutions. The series will serve as a multi-disciplinary resource linking renewable energy with society. The book series fulfils the rapidly growing worldwide interest in energy solutions. It covers all fields of renewable energy and their possible applications will be addressed not only from a technical point of view, but also from economic, social, political, and financial aspect.

Books published in the series include research monographs, edited volumes, handbooks and textbooks. The books provide professionals, researchers, educators, and advanced students in the field with an invaluable insight into the latest research and developments.

Topics covered in the series include, but are by no means restricted to the following:

- Renewable energy
- Energy Solutions
- Energy storage
- Sustainability
- Green technology

For a list of other books in this series, visit www.riverpublishers.com

Advanced Concepts for Renewable Energy Supply of Data Centres

Editors

Jaume Salom

IREC – Catalonia Institute for Energy Research
Spain

Thorsten Urbaneck

Technische Universität Chemnitz
Germany

Eduard Oró

IREC – Catalonia Institute for Energy Research
Spain

River Publishers

Routledge
Taylor & Francis Group
LONDON AND NEW YORK

Published 2017 by River Publishers

River Publishers
Alsbjergvej 10, 9260 Gistrup, Denmark
www.riverpublishers.com

Distributed exclusively by Routledge

4 Park Square, Milton Park, Abingdon, Oxon OX14 4RN
605 Third Avenue, New York, NY 10158

First published in paperback 2024

Advanced Concepts for Renewable Energy Supply of Data Centres / by Jaume Salom, Thorsten Urbaneck, Eduard Oró.

Routledge is an imprint of the Taylor & Francis Group, an informa business

Publisher's Note
The publisher has gone to great lengths to ensure the quality of this reprint but points out that some imperfections in the original copies may be apparent.

While every effort is made to provide dependable information, the publisher, authors, and editors cannot be held responsible for any errors or omissions.

ISBN: 978-87-93519-42-8 (hbk)
ISBN: 978-87-7004-425-7 (pbk)
ISBN: 978-1-003-33699-0 (ebk)

DOI: 10.1201/9781003336990

Contents

Eduard Oró and Jaume Salom

2 Operational Requirement   27

Eduard Oró, Victor Depoorter and Jaume Salom

3 Environmental and Economic Metrics for Data Centres   41

Jaume Salom and Albert Garcia

7 Applying Advanced Technical Concepts to Selected Scenarios 177

Verena Rudolf, Nirendra Lal Shrestha, Eduard Oró,
Thorsten Urbaneck and Jaume Salom

Preface

The rapid increase of cloud computing and the vast growth in Internet and Social Media use have aroused the interest in energy consumption and carbon footprint of Data Centre industry. Data Centres primarily contain information technology (IT) equipment used for data processing, data storage, and communications.

The book introduces a number of energy efficient measures and advanced solutions for the supply of power and cooling energy into Data Centres with the goal to achieve Net-Zero Energy Data Centres. Because of the high energy density nature of these unique infrastructures, it is essential to implement energy efficiency measures and reduce its consumption before introducing any renewable energy source. A holistic approach is implemented to integrate different strategies such as IT load management, efficient electrical supply, integration of Low-Ex cooling systems (i.e. free cooling), interaction with district heating and cooling networks, implementation of heat reuse solutions, and optimal use of thermal and electrical storage.

The book presents a catalogue of advanced technical concepts that could be integrated into Data Centres portfolio in order to increase the overall efficiency and the share of renewable energies in power and cooling supply. Based on dynamic energy models developed in TRNSYS the most promising solutions are deeply studied through yearly simulations. A set of environmental metrics (i.e. non-renewable primary energy), financial metrics (i.e. CAPEX, OPEX and TCO), and energy efficiency metrics (i.e. PUE), are described and used to evaluate the different technical concepts.

The book is organized in seven chapters. The first chapter introduces to the Data Centre industry, describing the basic structure of a Data Centre and its different typology of infrastructure, workloads and redundancy levels. Moreover, market segmentation and actual trends as well as the implementation of renewables in actual Data Centre industry are also showed. Chapter two presents the IT equipment requirements which are those requirements needed for a proper Data Centre operation (i.e. temperature and humidity ranges of operation). The third chapter describes the relevant metrics to evaluate the performance of different advanced energy concepts applied to Data Centres under a holistic view. Besides energy efficiency other aspects need to be considered to assess the overall environmental impact of Data Centres which are the use of renewable resources, the usage of deployed capacity and the economic impact. Chapter four introduces different strategies and concepts for the reduction of power and energy demand of Data Centres. First, strategies for efficient IT management which include consolidation policies and workload shifting are presented. Further, the chapter describes how the efficiency of the electric distribution can be increased to reduce the cooling and electric load, as for example the use of modular and bypass UPS. Chapter five describes current best practices and novel energy efficiency measures which can improve the supply and distribution of cooling in Data Centers. The sixth chapter presents six advanced technical concepts to supply the electrical and cooling load efficiently with a high share of renewable energy resources. A detailed description including thermal and electric schemes is presented for each concept. This comes along with the description of the main components of each concept and an analysis of the yearly energy flows using Sankey charts. Finally, chapter seven summarizes the energetic and economic performance of the concepts. Results of simulations over an entire year are used to investigate the main Data Centre's characteristics such as the location and the size of the Data Centre while comparing the concepts between them.

Advanced Concepts for Renewable Energy Supply of Data Centres is a book addressed to personnel from the Data Centre industry, both IT and facility engineers, as well to business managers. It is also addressed to academic staff and collegial, master and Ph.D. students in computer science, and mechanical and electrical engineering.

Acknowledgments

The current book is the final step to disseminate some of the outstanding results derived from the EU RenewIT project (www.renewit-project.eu). The project RenewIT – *Advanced concepts and tools for renewable energy supply of IT Data Centres* has received funding from the European Union's Seventh Framework Programme for research, technological development and demonstration under grant agreement no. 608679. The recommendations and opinions expressed in the book are those of the editors and contributors, and do not necessarily represent those of the European Commission.

First and foremost, we would like to express our warm appreciation to all the institutions and persons that have been participated in the RenewIT project and made key contributions which make possible to produce this book. Expertise of all the members in the RenewIT consortium has been very relevant to make a step ahead in this transdisciplinary project. They are:

- DEERNS NEDERLAND BV, The Netherlands
- SISTEMES AVANCATS DE ENERGIA SOLAR TERMICA SCCL – AIGUASOL, Spain
- TECHNISCHE UNIVERSITAET CHEMNITZ – TUC, Germany
- 451 RESEARCH (UK) LIMITED, United Kingdom
- AEA s.r.l. LOCCIONI Group, Italy
- BARCELONA SUPERCOMPUTING CENTER – CENTRO NACIONAL DE SUPERCOMPUTACION – BSC, Spain
- INSTITUT DE RECERCA EN ENERGIA DE CATALUNYA – IREC, Spain

Appreciated thanks go to our authors who contributed to produce excellent book chapters, especially to Verena Rudolf and Nirendra Lal Shrestha, who dedicated a huge effort in the final edition of the chapters five, six and seven. Beyond the authors, we would like to express our acknowledgement to persons who definitely contributed in the RenewIT project to produce the tools, models and knowledge which are on the basis of what is explained in the different chapters. They are Bianca Van der Ha, Bert Nagtegaal, Mieke Timmerman,

Hans Trapman, Gilbert de Nijs and Joris van Dorp (Deerns); Angel Carrera and Oscar Càmara (Aiguasol); Mario Macias and Jordi Guitart (BSC) and Andrew Donoghue (451 Research).

Our warm appreciation to the River Publisher staffs who allowed us to publish our work and gave their valuable time to produce and review our book. Also, we want to thank to the OpenAIRE – EC FP7 post-grant Open Access Pilot to provide funding for Open Access publishing.

Finally, we want to thank our families who supported and encouraged us in spite of all the time it took us away from them.

List of Contributors

Albert Garcia, *Catalonia Institute for Energy Research – IREC, Spain*

Eduard Oró, *Catalonia Institute for Energy Research – IREC, Spain*

Jaume Salom, *Catalonia Institute for Energy Research – IREC, Spain*

Mauro Canuto, *Barcelona Supercomputing Center – Centro Nacional de Supercomputación, Spain*

Nirendra Lal Shrestha, *Chemnitz University of Technology, Professorship Technical Thermodynamics, Germany*

Noah Pflugradt, *Chemnitz University of Technology, Professorship Technical Thermodynamics, Germany*

Thomas Oppelt, *Chemnitz University of Technology, Professorship Technical Thermodynamics, Germany*

Thorsten Urbaneck, *Chemnitz University of Technology, Professorship Technical Thermodynamics, Germany*

Verena Rudolf, *Chemnitz University of Technology, Professorship Technical Thermodynamics, Germany*

Victor Depoorter, *Catalonia Institute for Energy Research – IREC, Spain*

List of Figures

List of Tables

List of Symbols and Abbreviations

Latin Symbols

Symbol	Description	Unit
A	area	m^2
A_R	area of turbine blade	m^2
c	specific costs	€/kW
C	costs, concentrating factor, capacity flow	€
C_I	investment cost	€
$C_{a,i}$	annual costs for the year i	€
ce	cost in terms of energy capacity	€
$C_{e,DC}$	energy costs for DC	€
c_p	specific heat capacity at constant pressure	J/(kgK)
C_p	power coefficient	–
$CAPEX$	capital expenditure	€
CC	connection capacity credit	–
$CC_{fac,des}$	facility designed capacity credit	–
$CC_{fac,inst}$	facility installed capacity credit	–
$CC_{IT,des}$	IT designed capacity credit	–
$CC_{IT,inst}$	IT installed capacity credit	–
$CC_{IT,m2}$	white space capacity credit	–
CCF	cooling capacity factor	–
COP	coefficient of performance	–
DR	dimensioning rate	–
dt	increment of time	s
E, e	energy	J, kWh
E_{air}	energy delivered by air	J, kWh
E_{fuel}	energy delivered by fuel	J, kWh
EM_{DC,CO_2}	data centre CO_2 emissions	Kg CO_2
f_{RPI}	Mean annual rate of cost increases	–
f_t	correction factor	–
$fsav$	fractional savings	–

G	solar irradiance	W/m^2
GUF	grid utilization factor	–
h	specific enthalpy	J/kg
K_p	factor of proportional gain	–
LHV	lower heating value	J/kg
$\dot{m}$	mass flow	kg/s
$MSOC$	Minimal state of charge	–
$OPEX$	operating expenditure	€
p	pressure	Pa
P	Power, pump	W
ΔP	pressure difference	Pa
$PE_{DC,nren}$	non-renewable primary energy	kWh
$PE_{DC,tot}$	total primary energy	kWh
PIT	IT power	kW
Q	heat quantity	J, kWh
$\dot{Q}$	heat flow	W
$\dot{Q}_0$	design cooling capacity	W
$\dot{Q}_C$	cooling power	J, kWh
$\dot{Q}_{con}$	heat rejection rate	W
Q_H, Q_h	heat	J, kWh
R_d	discount rate	–
ROI	return of investment	€/time
S	Storage energy flow	kWh
t	time	s
T	temperature	K, °C
T_{cow}	cooling water temperature	K, °C
T_S	supply temperature	K, °C
$T_{db,d}$	design dry-bulb temperature	K, °C
ΔT	temperature difference	K
U_o	overall heat transfer coefficient	W/(m^{2}K)
v	wind velocity	m/s
$\dot{V}$	volume flow	m^3/s
$V_{f,\tau}$	Final value	€
W	electrical energy	J, kWh
w	weight factor	–
$Water_{DC}$	annual water consumption	m^3/a
y	year	a

Greek Symbols

Symbol	Description	Unit
ϕ	load factor	–
γ_{load}	load cover factor	–
γ_{supply}	supply cover factor	–
η	energy conversion efficiency	–
η_m	electric generator efficiency	–
η_t	gearbox efficiency	–
ρ	density	kg/m^3
ξ	thermal losses	J, kWh
ζ	electrical losses	kWh

Subscripts

a	air
Abs	absorber
amb	ambient
Aux, AUX	auxiliary
bat	battery
BOS	balance of system
C, c	cooling
CDS	charging/discharging system
chw	chilled water
CM	Cost of maintenance
CO_2	Greenhouse gas emissions
con	condenser
cow	cooling water
DC	Data Centre
dch	discharging
DCool	district cooling
del	delivered
des	design
DHeat	district heating
EC	energy costs
eff	effective
el	electricity
EP, PE	primary energy
ev	evaporator
exp	exported

f	fan, final
fac	facility
FC	fuel cell, free cooling
Fuel	fuel (i.e, natural gas, biogas, etc.)
g	global
G	grade
h, H, heat	heat
Hex, HEX	heat exchanger
Hp, HP	heat pump
hsi	heat sink
hw	hot water
i	energy carrier
imp	imported
in	inlet
inst	installed
IST	Ice storage tank
IT	information technology
itcw	IT cooling water
m	mechanical, mean
max	maximum
min	minimum
NOCT	nominal operating cell temperature
nom	nominal
nren	non-renewable energy
out	outlet
r	return
REC	replacement
ref	reference case
ren	renewable energy
s	supply
th	thermal
tot	total
ws	white space

Abbreviations

ABCH	absorption chiller
AC	alternating current
ACE	availability, capacity and efficiency

AFC	alkaline fuel cell
AGFW	German Heat and Power Association
AHU	auxiliary heating unit
ALM	Almeria
AMS	Amsterdam
AMWS	annual mean wind speed
ANC	Ancona
APP	average power price
ASHRAE	American Society of Heating, Refrigerating and Air-Conditioning Engineers
ATES	aquifer thermal energy storage
BCN	Barcelona
BEG	Belgrade
BGO	Bergen
BIPV	building integrated photovoltaic
BMS	building management system
CBS	cold buffer storage
CCF	cooling capacity factor
CDSP	cost driven scheduling policy
CHE	Chemnitz
CHP	combined heat and power
CHWST	chilled water storage
CO_2 savings	Avoided carbon emissions
CONV	converter
COP	coefficient of performance
CPU	computer processing unit
CRAC	computer room air conditioning
CRAH	computer room air handler
CT, CTW	cooling tower
CV	control valve
CWB	cooling water basin
DAFC	direct air free cooling
DC	district cooling, direct current, data centre
DCIM	Data Centre infrastructure management
DEERNS	Deerns Nederland B.V.
DH	district heating
DHW	domestic hot water
DIEC	direct evaporative cooler
DMFC	direct methanol fuel cell

DP	dew point
DR	dimensioning rate
DRC	dry cooler
DRUPS	diesel rotary uninterruptible power supply
DSE	digital service efficiency
DX	direct expansion
EC	energy carrier
EDI	Edinburgh
EE	energy efficiency
EER	energy efficiency ratio
EES	electrical energy storage
EPDM-HT	ethylene propylene diene monomer – high temperature
EPRI	Electric Power Research Institute
FC	free cooling, fuel cell
FRA	Frankfurt
GCDC	green cooling for data centre
GE	generator
GRO	Groningen
GUI	graphical user interface
HEX	heat exchanger
HDD	hard disk drive
HST	heat storage tank
HP	heat pump
HPC	high performance computing
HPL	high performance computing linpack benchmark
HS	hydraulic separator
HIS, hsi	heat sink
HVAC	heating, ventilation and air conditioning
HWST	hot water storage
HWT	horizontal axis wind turbine
IAFC	indirect air free cooling
ICT	information and communication technologies
IEEE	Institute of Electrical and Electronics Engineers
INN	Innsbruck
IREC	Catalonia Institute for Energy Research
ISP	internet service provider
IT	information technology
KBP	Kiev
KPI	key performance indicator

KUN	Kaunas
LON	London
M/G	motor/generator
MAD	Madrid
MCFC	molten carbonate fuel cell
MERV	minimum Efficiency Reporting Value
MIL	Milan
MISC	miscellaneous
MPI	message passing interface
MS	mechanical system
MTBF	mean time between failures
MTS	mechanical technical system
MTTR	mean time to repair
N	number
NREL	National Renewable Energy Laboratory
nZEB	nearly zero energy building
OAT	one-at-a-time
OPO	Porto
P	pump
PAR	Paris
PACF	phosphoric acid fuel cell
PE	primary energy
PEsavings	primary energy savings
PEM	proton exchange membrane
PDU	power distribution unit
PDTS	power distribution technical system
PECF	primary energy conversion factor
PEM	proton exchange membrane
PMSM	permanent magnet synchronous machine
PR	performance ratio
PSU	Power supply unit
PUE	power usage effectiveness
PV	photovoltaic
RAM	random access memory
REF	renewable energy factor
REHVA	Federation of European HVAC Associations
RER	renewable energy ratio
REDUN	redundancy

REGO	Renewable Energy Guarantees of Origin
RH	relative humidity
ROM	Roma
ROT	Rotterdam
SE	single effect
SEER	seasonal energy efficiency ratio
SLA	service-level agreement
SOC	state of charge
SOFC	solid oxide fuel cell
SPH	space heating
SRV	server
STK	Stockholm
STS	static transfer switch
SUB	substation
SUPS	static uninterruptible power supply
SVQ	Seville
SWAC	seawater air conditioning
SWG	switch gear
TCO	total cost of ownership
TES	thermal energy storage
TotalEPPercent	total renewable energy consumed
TR	transformer
TWV	three way valve
UPS	static uninterruptible power supply
VAL	Valencia
VAF	variable air flow
VCCH	vapour compression chiller
VER	Verona
VFD	variable-frequency drive
VM	virtual machine
VMM	virtual machine manager
VWT	vertical axis wind turbine
WAW	Warsaw
WCT	wet cooling tower
WSP	white space
WT	wind turbine
WUE	water usage effectiveness
ZRH	Zurich

1

Data Centre Overview

Eduard Oró and Jaume Salom

Catalonia Institute for Energy Research – IREC, Spain

1.1 Data Centre Infrastructure

1.1.1 Introduction

The rapid increase of cloud computing, high-performance computing (HPC) and the vast growth in Internet and Social Media use have aroused the interest in energy consumption and the carbon footprint of Data Centres. Data Centres primarily contain electronic equipment used for data processing (servers), data storage (storage equipment) and communications (network equipment). Collectively, this equipment processes, stores and transmits digital information and is known as information technology (IT) equipment. Data Centres also usually contain specialized power conversion and backup equipment to maintain reliable, high-quality power as well as environmental control equipment to maintain the proper temperature and humidity for the IT equipment [1]. Data Centres are commonly divided into 3 different spaces (Figure 1.1):

- **IT room or data hall**. The IT room is an environmentally controlled space that houses equipment and cabling directly related to compute and telecommunication systems which generate a considerable amount of heat in a small area. Moreover, the IT equipment is highly sensitive to temperature and humidity fluctuations, so a Data Centre must keep restricted power and cooling conditions for assuring the integrity and functionality of its hosted equipment. Thus, many manufactures call the IT room as whitespace.
- **Support area**. These areas are those where different systems such as power, cooling and telecommunications are located. Basically, these areas are electrical gear and uninterruptible power supply (UPS) control,

1

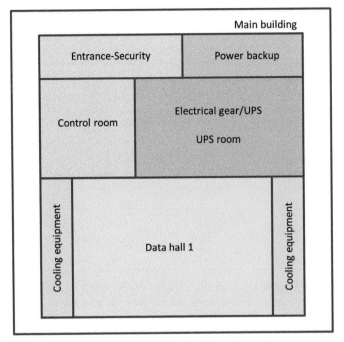

Figure 1.1   Scheme of a typical Data Centre layout.

UPS room, power backup system, cooling equipment areas, control room, etc.

- **Ancillary spaces**. These are those such as offices, lobby and restrooms.

1.2 Main Subsystems

Data Centres typically comprise three main subsystems: IT equipment, which provides services to customers; power infrastructure, which supports the IT and cooling equipment; and cooling infrastructure, which removes the generated heat by the IT and power equipment.

1.2.1 IT Equipment

IT equipment is the main contributor to electricity consumption of a Data Centre, representing about 45–55% of the total and can be described as the IT work capacity employed for a given IT power consumption [2]. Thus, the IT equipment power includes the load associated with all of the IT equipment, such as compute, storage and network equipment, along with supplemental equipment such as monitors, workstations/laptops used to monitor or otherwise control the Data Centres. Table 1.1 shows the IT equipment components and its classification.

1.2.2 Power System

The power distribution system of a Data Centre is intended as the equipment in charge of bringing electrical power to the loads of the system fulfilling adequate power quality levels and security of supply. In this sense, reliable power supply is one of the cornerstones of a successful Data Centre. Since the public grid could have short voltage dips which can cause servers to malfunction or even crash completely, it is essential to stabilize the power supply. Additionally, there is always the danger of longer power outages, which can crash the entire Data Centre. Figure 1.2 shows a scheme of the power supply for a standard Data Centre from the main grid to the IT equipment. Between the public grid and the infrastructure there is first the transformer which is responsible to transforms electrical energy from high/medium voltage to low voltage. After this electrical element, there is the switchgear, which is used to control, protect and isolate electrical equipment. In the framework of Data Centres, it is responsible to connect the IT equipment to the main grid or to the backup diesel generator. As mentioned, usually there is a backup diesel generator to supply power in case that the public grid presents a problem and the battery or flywheel is usually designed to be sufficient to keep the IT equipment running until the diesel generators can be started (few seconds). Moreover, there are other components that support the IT equipment such as the UPS, switchgear and power supply units (PDU), and other miscellaneous component loads such as lighting and filters.

Table 1.1   IT equipment components

Servers	Compute devices	Servers
Networking	Network devices	Switches
		Routers
		Printers
	IT support systems	PCs/workstations
		Remote management
	Miscellaneous devices	Security encryption, storage encryption, appliances, etc.
Storage	Storage	Storage devices – switches, storage array Backup devices – media libraries, virtual media libraries
	Telecommunication	All Telco devices

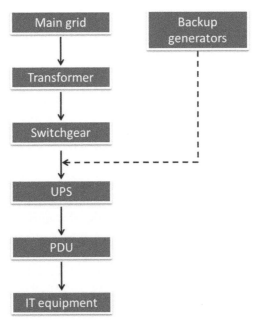

Figure 1.2   Scheme of the power supply in a standard Data Centre.

There are many different UPSs which can vary greatly in physical size, weight, storage capacity, supported input power, cost, etc. Here, the most known ones will be presented and described being divided between static and rotary UPS. The UPSs which do not have any moving parts throughout its power path are the so-called static UPS. On the other hand, when there are rotating components such as a motor generator within the UPS to transfer power to the load, they are called rotary UPSs. Another fundamental difference between them is that static UPS uses chemical stored energy (in the batteries) in the event of a power cut whereas rotary UPS uses the stored kinetic energy of a spinning flywheel. Rotary UPSs have relatively low redundancy (typically between 15 and 30 seconds) on their own compared to the static UPS. However, this low redundancy allows sufficient time for the generator to start up and accept the full load without affecting the output to the Data Centre. The efficiency performance of static UPSs is higher than for rotary ones, especially at partial loads. This means that the rotary UPSs sustain higher fixed losses such as the energy utilized to power controls, flywheels and pony motors associated with the rotary UPS at zero load and the energy utilized to preheat the engine coolant and lubrication. In some cases, at full load, the efficiency of the rotary UPS can be higher [3]. However, technology is moving faster and the efficiency of these elements is improving day by day; therefore, for each particular case, a comparison study between both technologies should be done.

Figure 1.3 shows different approaches for distributing power to IT racks seen in Data Centres today. Mainly, there are three kinds of power distribution: panel-board distribution, traditional PDU distribution and modular distribution. In panel distribution, the main Data Centre power is distributed to multiple wall-mounted panel boards. Wall mount panel boards are a very low-cost power distribution strategy, made up of parts that an electrician can acquire in days, not weeks, to implement quickly. In traditional power distribution systems, the main Data Centre power is distributed to multiple PDU throughout the IT space. Moreover, there are two main categories of traditional PDU systems: field-wired using power cables in cable trays or in flexible or rigid conduit, distributed beneath the raised floor or above the IT racks. Finally, in order to meet the modern IT requirements, alternative power distribution approaches are appearing in Data Centres. These approaches are more flexible, more manageable, more reliable and more efficient. Panel-board distribution and field-wired traditional PDU distribution systems are shown to be the best approach when low first cost is the highest priority, when the IT space has unique room constraints, and when IT changes are

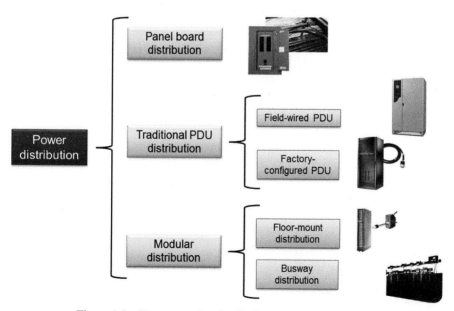

Figure 1.3   Five approaches for distributing power to IT racks.

not likely. Factory-configured traditional distribution systems are shown to be the best approach when a Data Centre requires portability of its equipment, when additional pods may be added in future, and when low first cost is still a priority. Modular distribution approaches allow for more flexibility, more manageability, increased reliability and increased efficiency, to better address the IT needs of many Data Centres today. Bus way is optimal when floor space is a constraint, when there is a large open floor plan with a well-defined IT equipment layout, and when there is a high confidence in the final capacity need. It is also ideal when there is a high frequency of IT equipment turnover requiring new circuits. Floor-mount modular distribution, on the other hand, is optimal when the Data Centre has an uncertain growth plan and locations are not precisely defined in advance, because it offers flexibility to place units in specific locations when the need arises. It is also best for Data Centres that are being retrofit with additional capacity (i.e. addition of a high-density zone) [4].

1.2.3 Cooling System

The physical environment inside the Data Centres is rigorously controlled, and therefore, the design of the cooling system is critical to the efficiency and reliability of the whole Data Centre. There are mainly two types of cooling: air-cooled systems and water-cooled systems.

The majority of existing Data Centres are refrigerated through air. The server racks are normally arranged in cold and hot aisle containments to improve the air management. In these systems, the chilled air produced by the CRAH unit is driven into the cold aisle, either through the floor plenum and perforated tiles or through diffusers in the ceilings. The warm air in hot aisles is captured and returned to the intake of the CRAH. The heat in the CRAH unit is absorbed by water into a chiller system. There is another air-cooling system which consists on in-row cooling, so the CRAH unit (and therefore the chilled water) is embedded next to the racks and the hot air does not flow through the whitespace but inside the containment. Figure 1.4 shows different kind of air cooled Data Centres: cold aisle containment, hot aisle containment and in-row cooling systems.

Some new Data Centre designs have power loadings to levels that are difficult to remove with CRAH units. Therefore, other cooling techniques, such as on-chip cooling (either single- or double-phase liquid systems) and submerged cooling systems are also used. When on-chip cooling is adopted, not all the heat from the server is absorbed by the liquid, and therefore,

Figure 1.4 Cold aisle containment, hot aisle containment and in-row cooling [5].

an auxiliary air-cooling system is also needed. The use of this technology raises a concern about leaks and can cause irrevocable damage if it comes into direct contact with IT equipment. This is an issue because maintenance, repair and replacement of electronic components result in the need to disconnect and reconnect the liquid carrying lines. To overcome this concern, it is possible to use nonconductive liquid such as refrigerant or a dielectric fluid in the cooling loop for the IT equipment. Thus, the processors can be directly immersed in these liquids, which improve thermal efficiencies and often results in simplified liquid cooling systems. Therefore, submerged cooling systems are able to absorb all the heat from the servers through a dielectric refrigerant. These cooling technologies can provide high performance while achieving high energy efficiency in power densities beyond air-cooled equipment and simultaneously enabling use of waste heat when supply facility water temperatures are high enough. Moreover, liquid cooling can also offer advantages compared to air in terms of heat exchange (since it has much higher heat conduction coefficient of 0.6 W/(m·K), lower noise levels and closer temperature control. Figure 1.5 shows a picture of each of the liquid cooled technologies available nowadays.

Figure 1.5 On-chip water cooling, on-chip two-phase cooling system and submerged cooling system [6].

1.3 Data Centre Archetypes

To provide a better overview of the Data Centre market, Data Centres are itemised based on the following characteristics:

- Function or objective of the Data Centre
- Size
- Location and surroundings

Figure 1.6 shows the distribution of a Data Centre in function of the business model, size and location and surroundings.

1.3.1 Function or Objective of the Data Centre

Based on the European Standard EN 50600-1:2012, there are three types of Data Centres: enterprise Data Centres, co-location Data Centres and hosting Data Centres. The main difference between these types of Data Centres is the owner of the Data Centre facility and the IT, and with that the possibility

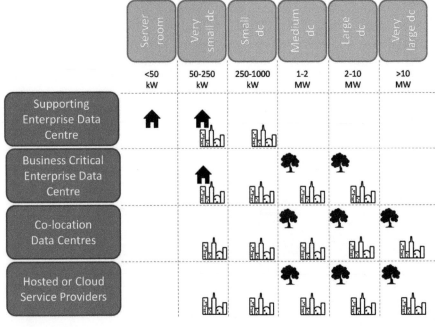

Figure 1.6 Overview of Common Data Centre Archetypes based on size (IT power), Data Centre function or objective and surrounding [7].

to synchronize the Data Centre systems and the IT. However, the enterprise Data Centre group can also be divided into two subgroups for a more clear differentiation: the business supporting Data Centres and the business critical Data Centres. Thus, this results in the following four main groups:

1. **Enterprise** Data Centres. These are built and operated by the end user, which are banks, hospitals and retailers, including HPC at universities or research institutions. For enterprises, a Data Centre can have two distinctive functions as will be described below.

   (a) **Business supporting** Data Centre. Firstly, the function of the Data Centre is to support the activities of the firm. For instance, a contractor uses a Data Centre to save and backup all the designs, drawings, 3D modelling activities and administration. Similarly, a university uses its Data Centre to perform among others, (high performance) research activities, backup and storage facilitation and administrative computing. In general, these Data Centres will provide safe, secure and reliable hosting facilities for the firms core IT systems. Since the Data Centres are not leading, but supporting, they are most frequently situated close to the actual firm or organisation, and therefore at short distance of the actual activities.

   (b) **Business Critical** Data Centre. They are an integral part of the main business process. These are, for example, the commercial telecom Data Centres and Data Centres of financial institutions. A bank uses its Data Centre for all the transactions, or telecom providers use the Data Centres for their telecom data transmissions. For these types of enterprises, the Data Centre is a leading necessity within their business process. Therefore, these Data Centres are situated at locations that are beneficial for the IT processes, based on criteria such as (not limited) distance to the customers, distance to a (large) power plant, cost and availability of land, (transatlantic) glass fibre connectivity or carrier neutrality options.

2. **Co-location** or commercial/multi-tenant Data Centres. In a co-location Data Centre multiple customers are housed with their own IT networks, servers and equipment. The co-location Data Centre provides the supporting infrastructures to its customers (mainly security, power and cooling).

3. Hosted or **Cloud** service providers. Third-party organisations built and operate these facilities. They provide direct computing services to other

businesses. This class of Data Centres is important as it includes some of the largest names in IT (Google, Microsoft, Amazon) and some of the largest facilities in terms of square meters and numbers of servers. For these Data Centres, the IT is completely leading, so when deciding the location, factors such as the energy supply, Internet connection and costs of the plot are leading.

1.3.2 Size

IT power capacity is commonly used as the way to express the size of the Data Centre. The following breakdown based on size is identified:

1.	Server room	<50 kW
2.	Very small Data Centre	50–250 kW
3.	Small Data Centre	250–1000 kW
4.	Medium size Data Centre	1–2 MW
5.	Large Data Centre	2–10 MW
6.	Very large Data Centre	>10 MW

1.3.3 Location and Surroundings

The location and the surroundings of the Data Centre have influence on the possibilities of collaborating with its surroundings. When a Data Centre is close to offices, this creates more possibilities for these offices to use the heat of the Data Centre. However, when a Data Centre is not surrounded by any buildings, there is generally more space and less regulatory conditions for large-scale application of renewables such as windmills and solar panels. When looking at location and surroundings, the following parameters are distinguished:

- **Integrated** in another building (e.g. office)
- **Not integrated** in another building:
  - **Surrounded** by other buildings
  - **Stand alone**

1.3.4 Archetypes Definition

Based on this overview of Data Centre characteristics, the following archetypes of the Data Centres are detailed. Figure 1.7 shows the specified Data Centre archetypes. These are further specified in the text below:

1. Very small (less than 250 kW) Data Centre or server room inside a building. The Data Centre supports the activities of the enterprise.
2. Very small (less than 250 kW) Data Centre in a separate building surrounded by other buildings. The Data Centre supports the activities of the enterprise.
3. Small (between 50 and 1000 kW) Data Centre inside a building or in a separate building surrounded by buildings. The Data Centre is critical to the business.
4. Medium to large size (between 1 and 10 MW) Data Centre completely stand-alone. The Data Centre is critical to the business.
5. Medium to large size (between 1 and 10 MW) Data Centre in a separate building surrounded by other buildings. The Data Centre is critical to the business.
6. Small (between 50 and 1000 kW) co-location or hosting Data Centre surrounded by other buildings.
7. Medium to very large size (between 1 and >10 MW) stand-alone co-location or hosting Data Centre.

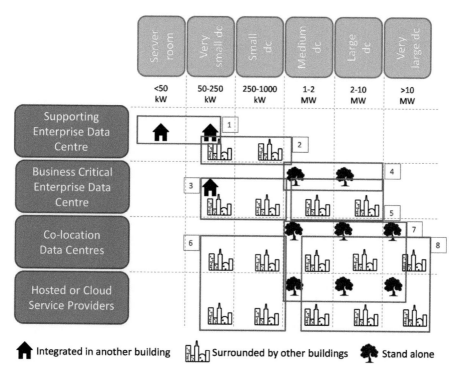

Figure 1.7 Data Centre archetypes defined by RenewIT project [7].

8. Medium to very large size (between 1 and >10 MW) co-location or hosting Data Centre surrounded by other buildings.

1.4 Workload Typology

The power consumption of a Data Centre depends on the current computing load and therefore the usage patterns of the IT equipment. In real Data Centres, there are disparities in performance and power characteristics across servers and different scheduling, task migration or load balancing mechanisms have effects on the IT power consumption and therefore cooling consumption. For simplicity, and according to the service that each Data Centre type provides, three main homogeneous IT workloads can be identified: web, HPC and data workload.

1.4.1 Web Workloads

Web workload has real-time requirements, the users of such workload need to get a response to their petitions in few seconds (i.e. Google search, Facebook surf, etc.). Therefore, the evaluation of web application resource consumption requires realistic workload simulations to obtain accurate results and conclusions. Although networking workloads are very heterogeneous and they change dramatically over relatively short time frames, they also have general traits that are common to all workloads, such as the daily cycle of activity. Characterizing the workload on the worldwide web has two components: the data available on servers and how clients access that data. There is no typical resource consumption profile for web workloads but they may use CPU, memory, network or disk in several proportions. In order to make a complete characterization of both client workloads and resource consumption, real production logs were taken from a top Online Travel Agency (OTA) in [8], observed in the 35+ physical node cluster in which the application was deployed. Figure 1.8 shows two traffic patterns for a dataset, including number of hits relative to peak load (a) and number of user sessions started (b) over one week, grouped in 30-minute periods. As it can be observed, the traffic decreases over the night, until it starts growing again soon in the morning. It keeps growing until noon, when it slightly decreases. Finally, the workload intensity starts increasing again over the afternoon until it reaches its maximum around 9:00 pm. Over the night, the traffic volume decreases until it finally reaches the beginning of the cycle again. Therefore, web workload has the particularity to follow a daily/weekly pattern.

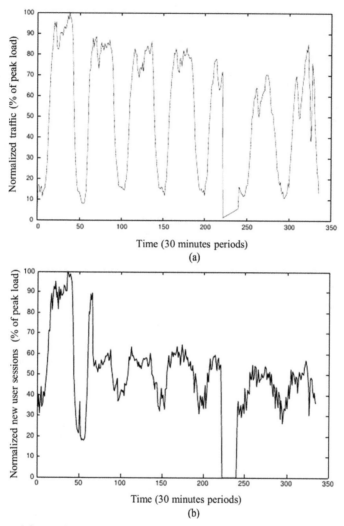

Figure 1.8 Typical profile for the traffic volume intensity during one week [8].

1.4.2 HPC Workloads

HPC workload is typically CPU intensive. They perform a large amount of floating-point operations for scientific calculations. Because HPC workloads may last for hours, or even days, they do not have real-time requirements, and they are usually allocated in job queues that may execute them hours or days after they are submitted by the users. Figure 1.9 shows power profiles of

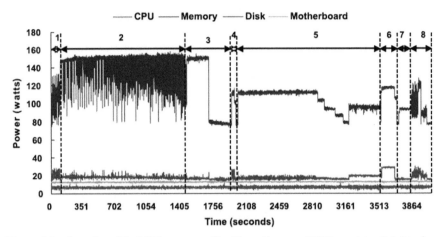

Figure 1.9   Snapshot of the HPC power profile. The entire run of HPC consists of eight micro benchmark tests in this order: (1) PTRANS; (2) HPL; (3) Star DGEMM + single DGEMM; (4) Star STREAM; (5) MPI RandomAccess; (6) Star RandomAccess; (7) Single RandomAccess; (8) MPI FFT, Star FFT, single FFT and latency/bandwidth [8].

the HPC benchmarks in a cluster made up of nine dual-core, Opteron-based server nodes. The power consumption of major components per compute node was obtained when running a full HPC benchmark suite using eight cores. These profiles are obtained using the problem size where HPL achieves its maximum performance on two nodes. In this test, the problem size fits the peak execution rate for HPL. For LOCAL tests, a benchmark on a single core with three idle cores was performed. Power consumption is tracked for major computing components including CPU, memory, disk and motherboard. These four components capture nearly all the dynamic power usage of the system that is dependent on the application. Notice that HPC workloads do not follow a given pattern, and they will depend on the access policy and dimension of the Data Centres of each institution.

1.4.3 Data Workloads

Data workload is usually both memory and disk intensive, while they can also use a high rate of CPU operations for data analysis. Despite the fact that data workloads may have real-time requirements (i.e. a search query in Google), data workloads can also have no real-time requirements (i.e. background data analytics for business intelligence applications). Obtaining accurate information on disk drive workloads is a complex task, because disk drives are deployed in a wide range of systems and support a highly diverse set of applications. The dynamics of file access in distributed systems are interesting in the sense that caching may cause the characteristics of the workload to look different at the client and at the server level. For example, locality is reduced near the servers because caches filter out repetitions, while merging of different request streams causes interference [9]. Notice that data workloads do not follow a given pattern, and they will depend on the access policy and dimension of the Data Centres of each institution.

1.4.4 Consumption versus Workload Typology

In order to predict the Data Centre consumption from the IT load, a relationship between server usage (IT load) and server consumption is needed. Firstly, it is needed to define a proper workload. In the framework of the RenewIT project, three homogeneous workloads were studied. Web workload is a real pattern collected from the access log of an ISP within the UPC [10], while HPC and data workloads patterns are extracted from the CEA-Curie Data Centre which are publicly available in the Parallel Workloads Archive [11]. Figure 1.10 presents the three homogeneous IT load profiles.

Secondly, the definition of IT load is an additive function that considers the load rates of CPU, main memory, disk and network, pondered according to the measured impact of each term in late 2012 servers [13]. Firstly, different types of micro-benchmark for fully stressing the system were executed in

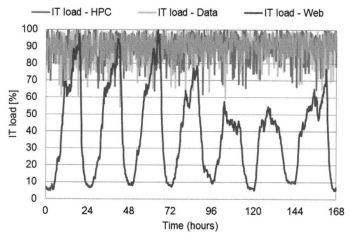

Figure 1.10   Different IT workload profiles (web, HPC and data) during a week [12].

order to reach the maximum real power of the system. These benchmarks included Ibench suite [14], Stress-ng [15], Sysbench [16], Prime95 [17], Pmbw [18], Fio [19] and Iperf3 [20]. After this initial process, different benchmarks based on real-world software stacks from CloudSuite [21] for web and data benchmarks, and NAS Parallel Benchmarks [22] for HPC were also executed.

With this experimentation, the relation between IT load and power consumption can be drawn for different server architecture. Notice that for its further adaptation to other hardware, these correlations were normalized. Figure 1.11 shows the results of the experimentation and the regressions to predict different consumptions in function of the IT load. The y-axis shows the percentage of the power referred to the maximum detected. The x-axis shows the percentage of the IT load referred to the maximum executed.

The variability in the power/load measurements shows that there is not a generic power profile for software, because all the components of a host (CPU, memory, disk and network) do not work independently. They must be coordinated because there are dependencies between data and procedures (and the usage of resources is variable across the same execution, depending of the availability of their required inputs at a given time). Finally, Figure 1.12 shows the results from RenewIT project compared to a standard function [23]. Notice that at low IT load, the power consumption is similar between the standard regression and the ones found in the project. However, at higher loads, especially above 80% of the IT load the difference becomes important. This means that while some researchers or Data Centre operators assume that at 100% of the IT load, for instance 100% CPU usage, the server consumes 100% of the nominal power, the reality is different.

1.5 Redundancy Level

1.5.1 Basic Definitions

Redundancy is an operational requirement of the Data Centre that refers to the duplication of certain components or functions of a system so that if they fail or need to be taken down for maintenance, others can take over. Redundant components can exist in any Data Centre system, including cabling, servers, switches, fans, power and cooling. It is often based on the "N" approach, where "N" is the base load or number of components needed to function. N+1 means having one more component than is actually needed to function, 2N means having double the amount of total components, and 2N+1 is having

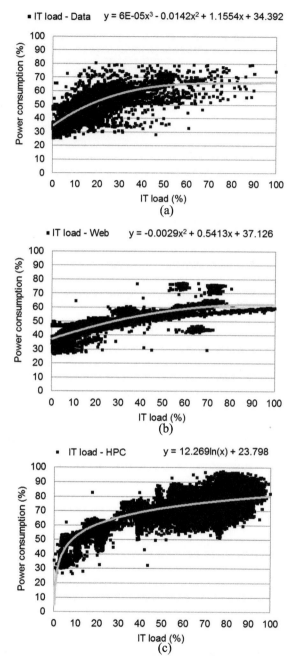

Figure 1.11 Curves for (a) data, (b) web and (c) HPC workload and their corresponding regression. Note that y means power consumption while x IT load in %.

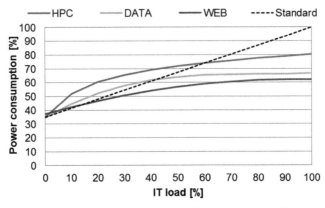

Figure 1.12   IT load relationship with power consumption.

double the amount plus one. While redundancy can be viewed as a "planned" operational function, availability is based on "unplanned" downtime and relates specifically to how many minutes or hours can be tolerated. For example, with 8,760 hours in a year, an availability of 99.9% indicates the ability to tolerate 8.76 hours of downtime per year. If the estimated cost of downtime within a specified time unit exceeds the amortized capital costs and operational expenses, a higher level of availability should be factored into the Data Centre design. If the cost of avoiding downtime greatly exceeds the cost of downtime itself, a lower level of availability should be factored into the design. Finally, reliability is the combination of redundancy and availability that go into determining reliability, and there are a variety of industry ratings that indicate reliability: Uptime Institute, TIA-942 and BICSI-002-2010. The Uptime Institute's Tier rating is the most common standard used and therefore it is described below.

1.5.2 Tier Levels

The Uptime Institute [24] defines a system of "Tiers" based upon business objectives and acceptable downtime which determines the redundancy of energy and cooling equipment. Moreover, TIERS levels can be described as follows:

- **Tier I**: Basic site infrastructure (N). It has non-redundant capacity components (components for cooling and electrical storage) and a single, non-redundant distribution path serving the computer equipment. It is estimated an availability of ~99.671% *and 28.8 hours of downtime per year*.

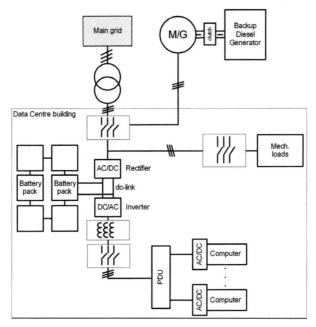

Figure 1.13 Power distribution system for Tier I considering static UPS (battery pack) [25].

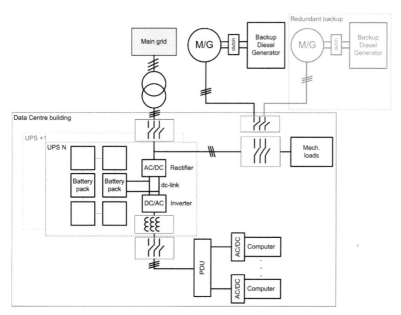

Figure 1.14 Power distribution system for Tier II considering static UPS [25].

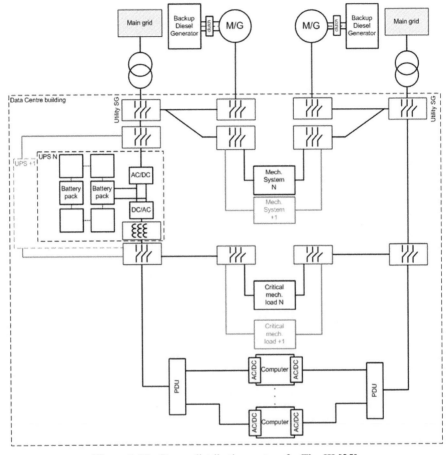

Figure 1.15   Power distribution system for Tier III [25].

- **Tier II**: Redundant site infrastructure capacity components (N+1). It has redundant capacity components and a single, non-redundant distribution path serving the computer equipment. It is estimated an availability of ~99.741% and *22 hours of downtime per year*.
- **Tier III**: Concurrently maintainable site infrastructure (2N). It has redundant capacity components and multiple independent distribution paths serving the computer equipment. Only one distribution path is required to serve the computer equipment at any time. It is estimated an availability of ~99.982% and *1.6 hours of downtime per year*.

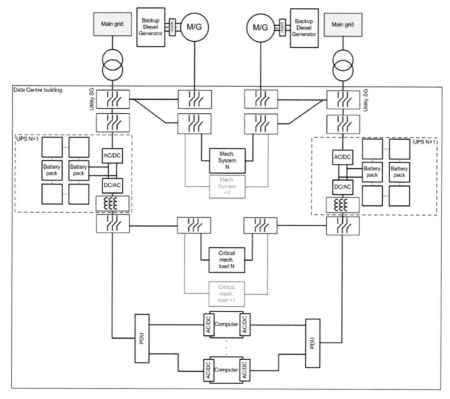

Figure 1.16 Power distribution system for Tier IV [25].

- **Tier IV**: Fault tolerant site infrastructure 2(N+1). It has multiple, independent, physically isolated systems that provide redundant capacity components and multiple, independent, diverse, active distribution paths simultaneously serving the computer equipment. The redundant capacity components and diverse distribution paths shall be configured such that "N" capacity providing power and cooling to the computer equipment after any infrastructure failure. It is estimated an availability of ~99.995% with 0.4 hours of downtime per year.

In general terms, Tier I corresponds to the basic structure and has non-redundant capacity components and distribution paths. Tier II is quite similar to Tier I, just including redundancies in capacity components but not in power distribution paths. Tiers III and IV contain redundancies both in capacity components and distribution paths, with Tier IV being the only fault tolerant site infrastructure. Therefore, failure or maintenance of any component will not affect computer equipment. Attending to this brief description of Tier levels classifying power distribution systems of Data Centres, electrical schemes for Tier I, Tier II, Tier III and Tier IV are presented in Figures 1.13–1.16, respectively.

1.6 Future Trends

Global power demand for Data Centres grew to an estimated 40GW in 2013, an increase of 7% over 2012 [26]. Most experts agree that this figure will continue rising in future despite improvements in IT performance as well as Data Centre design and operation. However, a recent report from NREL estimates almost no increasing in the United States Data Centre industry electricity consumption. These predictions always should be taken carefully since it is really difficult to predict the future trends. What for sure will increase is the Data Centre demand due to the arrival of Internet of Things, Internet and Social Media increasing demand from emerging countries.

Even though the most common Data Centres are small company-owned facilities, cloud computing has been driving major changes across the entire IT industry for nearly a decade, and its impact has been accelerating with every passing quarter. Many consultants predict that the impact of cloud and the growing role of hyper-scale operators are existential and severe and more work will go to co-location, hosting and cloud companies. That means there will be more IT and Data Centres in proportionately fewer hands. Figure 1.17 illustrates how capacity is shifting in terms of Data Centre space.

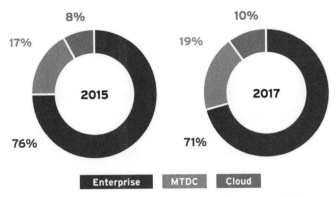

Figure 1.17   Global Data Centre space distribution [26].

A small but growing number of influential Data Centre owners and operators have seen financial, sustainability and availability benefits from increasing the proportion of energy they use from on-site and grid-based renewables such as wind, solar, hydropower and biofuels. On one hand, the average Data Centre is still likely to only derive a small percentage of its energy from grid or on-site renewables, but a handful of large Data Centres companies are skewing that trend. On the other hand, big players such as Google, Apple, Facebook and eBay have a track record of adopting disruptive approaches to Data Centre design and operation. Investment in renewable energy is another example of this disruptive approach in action.

Data Centre use of renewables can be broadly categorized as either from the grid or on-site. However, the reality is more nuanced, with a variety of

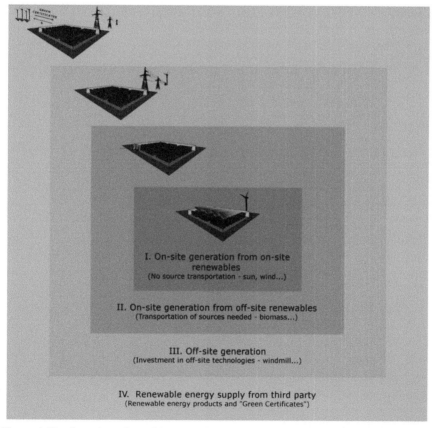

Figure 1.18   Overview of possible renewable supply options for Data Centre industry [27].

different approaches used to bring renewable energy into facilities. These range from generating renewables directly on-site to purchasing renewable energy credits from grid suppliers (see Figure 1.18). These different approaches can be broken down into four main groups:

- **On-site generation from on-site renewables**: Renewable energy is generated directly on-site through the use of solar panels or wind turbines.
- **On-site generation from off-site renewables**: Renewable energy is generated on-site but the fuel such as biomass, natural gas or biogas has to be transported to the facility.
- **Off-site generation**: Renewable energy is purchased from a nearby renewable energy provider. Technically, this could also be considered on-site generation depending on the proximity of the renewable generation source.
- **Renewables from third parties**: This includes purchasing Renewable Energy Guarantees of Origin (REGOs) in Europe or the US equivalent Renewable Energy Certificates (RECs). Other mechanisms include Power Purchase Agreements (PPAs) and Renewable Energy Tariffs (RETs).

References

[1] Shehabi, A. e. a. (2016). United States data center energy usage report. Ernest Orlando Lawrence Berkeley National Laboratory, Berkley, Californig. LBNL-1005775.

[2] IBM, Rear door exchanger planning guide, 2005–2008.

[3] Cottuli, C. Comparison of static and rotary UPS, White Paper 92. Schneider Electrics.

[4] Rasmussen, N., and Torell, W. (2013). Comparing Data Center Power Distribution Architectures. White Paper 129. Schneider Electric, 2013.

[5] Belden, "Belden," [Online]. Available: http://www.belden.com/blog/data centers/To-HAC-or-CAC-That-is-the-Question.cfm.

[6] "PHYS.ORG," [Online]. Available: http://phys.org/news/2012-09-intel-math-oil-dunk-cooler-servers.html.

[7] Van der Ha, B.N.B. (2014). Deliverable D7.1 Data Centres: AMrket Archetypes and Case Studies.

[8] Poggi, N., Moreno, T., Berral, J., Gavalda, R., and Torres, J. (2009). Self-adaptive utility-based web session management. *Comput. Networ.*, 53, 1712–1721.

[9] Froese, K.W., and Bunt, R.B. (1996). The effect of client caching on file server workloads, in *29th Hawaii International Conference System Sciences*.

[10] Macías, M., and Guitart, J. (2014). SLA negotiation and enforcement policies for revenue maximization and client classification in cloud providers. *Future Gener. Comp. Sy.*, 41, 19–31.

[11] [Online]. Available: http://www.cs.huji.ac.il/labs/parallel/workload/.

[12] Carbó, A. O. E. S. J. C. M. M. M. G. J. (2016). Experimental and numerical analysis for potential heat reuse in liquid cooled data centres. *Energy Conver. Manage.*, 112, 135–145.

[13] [Online]. Available: http://www.morganclaypool.com/doi/pdfplus/10.2200/S00516ED2V01Y201306CAC024.

[14] Delimitrou, C., and Kozyrakis, C. (2013). iBench: Quantifying interference for datacenter applications, in *2013 IEEE International Symposium on Workload Characterization (IISWC)*.

[15] [Online]. Available: http://kernel.ubuntu.com/~cking/stress-ng/.

[16] [Online]. Available: https://launchpad.net/sysbench.

[17] [Online]. Available: http://www.mersenne.org/download/.

[18] "Pmbw – Parallel Memory Bandwidth benchmark/measurement," [Online]. Available: http://panthema.net/2013/pmbw/.

[19] "Fio – Flexible I/O tester," [Online]. Available: http://git.kernel.dk/?p=fio.git;a=summary.

[20] "Iperf3: A TCP, UDP, and SCTP network bandwidth measurement tool," [Online]. Available: https://github.com/esnet/iperf.

[21] [Online]. Available: http://parsa.epfl.ch/cloudsuite/cloudsuite.html.

[22] [Online]. Available: http://www.nas.nasa.gov/publications/npb.html.

[23] Oró, E. (2014). Deliverable D4.2 Energy requirements for IT equipment, 2014.

[24] "Tier Classifications Define Site Infrastructure Performance," Uptime Institute.

[25] Shresthe, N. e. a. (2015). Deliverable D4.5 Catalogue of advanced technical concepts for Net Zero Energy Data Centres, 2015.

[26] Donoghue, A. (2015). Energizing renewable-powered datacenters. 451 Research, London.

[27] Oró, E. D. V. G. A. S. J. (2015). Energy efficiency and renewable energy integration in data centres. Strategies and modelling review. *Ren. Sustain. Energy Rev.* 42, 429–445.

[28] "Thermal guidelines for Data Processing environments," ASHRAE Datacom Series 1 – Third Edition.

[29] ASHRAE, "Gaseous and particulate contamination guidelines for data centres, Technical Commitee (TC) 9.9," ASHRAE, 2009.

[30] ASHRAE, "Gaseous and particle contamination guidelines for data centers.," 2011.

[31] ASHRAE, "Save energy now presentation series," 2009.

[32] Hintemann, R., Fichter, K., and Stobbe, L. (2010). Materialbestand der Rechenzentren in Deutschland. Umweltbundesamt, Dessau-Roßlau.

[33] Song, S., Ge, R., Feng, X., and Cameron, K. (2009). Energy profiling and analysis of the HPC challenge benchmarks. *International Journal of High Performance Computing Applications*, Vol. 23, pp. 265–276.

[34] Ren, Z., Xu, X., Wan, J., Shi, W., and Zhou, M. (2012). Workload characterization on a production Hadoop cluster: A case study on Taobao. in *IEEE International Symposium on Workload Characterizatio*.

[35] P. V. o. S. S. D. u. IBM, "http://www-03.ibm.com/systems/resources/ssd_ibmi.pdf," 2009. [Online].

[36] Gmach, D., Rolia, J., Cherkasova, L., and Kemper, A. (2007). Workload analysis and demand prediction of enterprise data center applications, in *IEEE 10th International Symposium on Workload Characterization*, Washington, DC, USA.

[37] Chen, Y., Alspaugh, S., and Katz, R. (2012). Interactive analytical processing in big data systems: A cross-industry study of MapReduce workloads. *PVLDB*, 5, 1802–1813.

[38] Ren, K., Kwon, Y., Balazinska, M. and Howe, B. (2013). Hadoop's adolescence. *PVLDB*, 6, 853–864.

[39] 16 December 2013. [Online]. Available: http://parsa.epfl.ch/cloudsuite.

[40] University of Hertfordshire, "TIA 942. Data Centre Standards Overview," 2014. [Online]. Available: http://www.herts.ac.uk/_data/assets/pdf_file/0017/45350/data-centre-standards.pdf.

2

Operational Requirement

Eduard Oró, Victor Depoorter and Jaume Salom

Catalonia Institute for Energy Research – IREC, Spain

2.1 Working Temperature Limit

2.1.1 Impact of Server Inlet Temperature

The rack/server supply air/water has an influence on the IT power consumption and also on the cooling system consumption. On one hand, the influence of temperature on the IT power consumption varies depending on the cooling system (air cooling or liquid cooling) and server architecture (i.e. AMD, Intel). There are several analyses available in literature characterizing this phenomenon. Figure 2.1 shows different correlations for liquid- and air-cooled systems [1]. As expected, the increment of supply air or water temperature increases the IT power consumption. Notice that the increments found in air cooling servers are higher than in the liquid, due to the fact that not only the current leakage is affecting the increase in the server consumption, but in the case of air cooled servers, the internal fans are responsible for this extra increase. The velocity of these fans normally is controlled internally by the server, and it is a function of the CPU temperature. Moreover, the life expectancy of computers goes down if the temperatures are too high, so the CPU manufacturers always specify maximum permitted temperatures for each CPU. On top of that, the server manufacturers specify maximum permitted temperatures for the entire server that is typically far lower than the permitted CPU temperatures.

On the other hand, working at high inlet, air/water temperatures reduces the energy consumption of the cooling system. For instance, if the inlet air temperature is 27°C in a Data Centre which also has an air-free cooling, most of the year the Data Centre will operate using free cooling and therefore minimizing the hours of mechanical cooling. Similarly, if the heat generated

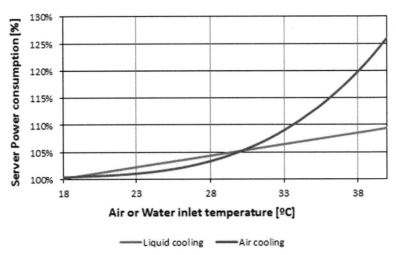

Figure 2.1   Variation of the IT power consumption for different cooling technologies [2].

by the IT equipment will be reused, the usefulness of the waste heat is much higher if the temperatures of the return water/air are higher.

Therefore, the goal is to find the optimal temperature range where the combined IT and cooling load is minimized. This temperature sweet spot varies by IT equipment, refrigeration technology, containment solution, the use of additional thermal energy efficiency strategies and other factors. It is important to note that not all Data Centres can immediately take advantage of the increase in operational temperatures. Historically, whitespace analysis results in a disparity in delivered temperatures due to inefficient air management systems, and therefore, an attempt to raise the temperature can put the systems located in hot spots at risk.

2.1.2 Permitted Temperatures of Individual Components

It is important to know that servers are built from many different systems such as memory RAM, CPU, HDD and main board. To gain a better understanding of the permitted temperatures of the individual components, Figure 2.2 shows the data found in the literature which can vary ±10°C [2]. Notice that this data can change in the near future since the IT technology is improving faster but to have the numbers in mind it is a good approximation. But the permitted temperatures are only part of the thermal requirements. More important is the influence of higher temperatures on the life expectancy. Figure 2.3 contains a

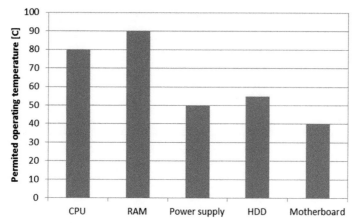

Figure 2.2 Permitted operating temperatures of different components.

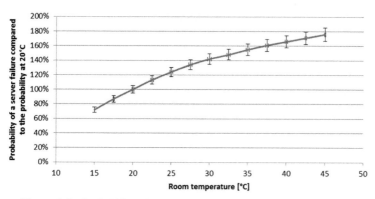

Figure 2.3 Probability of server failure versus room temperatures [3].

diagram based on data from [3] which shows the probability of a server failure compared to the probability at 20°C in function of the room temperature.

2.1.3 CPU Power Management and Throttling

Since recently, it used to be that the CPU was running at a certain frequency, and as long as the maximum permitted CPU frequency was not reached, it will be continue running at maximum speed. However, in the last few years, this management changed completely. To study deeply this phenomenon, a power management block was introduced in a processor which measures the temperatures and the power consumption trying to overclock it as much as possible based on the following criteria [4]:

- Number of active cores
- Estimated current consumption
- Estimated power consumption
- Processor temperature

The experimentation was done using a thermal budget power management in which the CPU was used as much as possible while ensuring not overheat of itself. Thus, the lower the CPU surface temperature is, the better the chance of actually using the entire thermal budget and the better the chance of getting the maximum possible performance from the CPU. So basically if a processor is requested to do a calculation and is cool enough, it will be over clocked and therefore finish the calculation significantly faster than at regular clock speed.

In air-cooled systems, there is a rather steep temperature gradient due to the limits of heat transfer between the CPU die and the cooling air flow. The die temperature at hot spots on the CPU can easily be 30–50 K higher than the air exit temperature, depending on air volume flow, heat sink design and other parameters. From [4] and in particular for an Intel Xeon E5-2697 v2, if the air temperature before the CPU is at 30°C and after the die at 40°C, the die temperature might be 60–70°C and since the maximum permitted CPU temperatures is 86°C, there is enough thermal budget to overclock. On the other hand, if the room temperature is 35°C, the air temperature at the CPU might be 45°C before the die and 55°C after the die, and then, the CPU would not over clock and performance would be lower. Also, the motherboard life expectancy might be in danger because important parts are "cooled" with 55°C air or more [5]. This phenomenon will obviously depend on the internal server configuration, and if there is some other IT equipment after the CPU or not.

2.2 Environmental Conditions

2.2.1 Temperature and Humidity Requirements

The IT equipment is the main contributor to electricity consumption and heat production of a Data Centre. Thus, this incorporates the entire load associated with the IT equipment, including compute, storage and network equipment, along with supplemental equipment such as monitors, workstations/laptops used to monitor or otherwise control the Data Centre. Traditionally, these unique infrastructures have had very controlled environments due to its singularity. In its thermal guidelines for data processing environments, summarized in Table 2.1, the ASHRAE [3] provides suitable environmental conditions for electronic equipment. Moreover, Figure 2.4 shows the temperatures and relative humidity recommended by the ASHRAE for all equipment classes. These values refer to the air inlet conditions in the IT equipment and thus into the room or the cold aisles in cold/hot aisles configuration. A bad control of humidity ranges can put at risk the reliability of the computing equipment. Very high humidity can cause water vapour to condensate on the equipment, while very low humidity can cause electrostatic discharges. Thus, ASHRAE recommends a humidity envelope between 20 and 80%. Besides temperature and humidity, air pollution could also cause failures in IT equipment.

Table 2.1 Summary of 2011 ASHRAE thermal guidelines for Data Centres [3]

	Dry-Bulb Temperature	Humidity Range	Maximum Dew Point
	Recommended		
Class A1 and A4	18–27°C	5.5°C DP to 60% RH and 15°C DP	–
	Allowable		
Class A1	15–32°C	20% to 80%	17°C
Class A2	10–35°C	20% to 80%	21°C
Class A3	5–40°C	8% to 85%	24°C
Class A4	5–45°C	8% to 90%	24°C

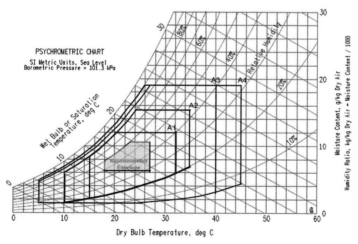

Figure 2.4 ASHRAE thermal guides for Data Centre operating environments [3].

2.2.2 Quality of the Room Air

Besides temperature and humidity, air pollution could also cause failures in Data Centres equipment. The effects of gaseous pollution and particles on different types of equipment failures are well documented. It is well known that moisture is necessary for metals to corrode but pollution aggravates it. Two common modes of IT equipment failures due to environmental contaminations are as follows [6]:

- Copper creep corrosion on printed circuit boards
- Corrosion of silver termination in miniature surface-mounted components.

Therefore, when operating in polluted geographies, Data Centre operators must also consider particulate and gaseous contamination that can influence the acceptable temperature and humidity limits within which Data Centres must operate to keep corrosion-related hardware failures rates at acceptable levels. Particulate (dust) contamination is characterized by its quantity and corrosiveness. The quantity of dust contamination can normally be identified by visual inspection of the IT equipment and by the filter replacement frequency. ISO 14644-1 has become the dominant, worldwide standard for classifying the cleanliness of air in terms of concentration of airborne particles. Table 2.2 provides maximum concentration levels for each ISO class. ASHRAE recommends that Data Centres be kept clean to ISO Class 8, which may be achieved simply by specifying the following means of filtrations:

- The room air may be continuously filtered with MERV 8 filters, as recommended by ASHRAE Standard 127 (ASHRAE 2007).
- Air entering a Data Centre may be filtered with MERV 11 or MERV 13 filters as recommended by ASHRAE (2009b).

For Data Centres utilizing free air cooling or air-side economizers, the choice of filters to achieve ISO class 8 level of cleanliness depends on the specific conditions presented at that Data Centre. In general, air entering a Data Centre may require use of MERV 11 or, preferably, MERV 13 filters.

Direct measurement of gaseous contamination levels is difficult and is not a useful indicator of the suitability of the environment for IT equipment. A low-cost, simple approach to monitoring the air quality in a Data Centre is to expose cooper and silver foil coupons for 30 days followed by coulometric reduction analysis in a laboratory to determine the thickness of the corrosion products on the metal coupons. ASHRAE recommends that Data Centre operators maintain an environment with corrosion rates within the following guidelines:

Table 2.2 ISO 14644-1 (ISO 1999) Air cleanliness classification vs maximum particle concentrations allowed [6]

ISO CLASS	Maximum Number of Particles in Air (Particles in Each Cubic Meter Equal to or Greater Than the Specified Size) Particle Size, μm					
	>0.1	>0.2	>0.3	>0.5	>1	>5
Class 1	10	2				
Class 2	100	24	10	4		
Class 3	1000	237	102	35	8	
Class 4	10,000	2370	1020	352	83	
Class 5	100,000	23,700	10,200	3520	832	29
Class 6	1,000,000	237,000	102,000	35,200	8320	293
Class 7				352,000	83,200	2930
Class 8				3,520,000	832,000	29,300
Class 9					8,320,000	293,000

- Copper reactivity rate of less than 300 A/month
- Silver reactivity rate of less than 200 A/month

For Data Centres with higher gaseous contamination levels, gas-phase filtrations which are commercially available are highly recommended.

2.3 Power Quality

Power quality is a concern when IT equipment has to work properly. As defined by international standards, there are five basic requirements related to power supply to support IT equipment without malfunction or damage which are described below.

2.3.1 Input Voltage within Acceptable Limits

Most equipment manufacturers use universal PSUs that can support the various input voltages and frequencies found worldwide. That means the PSU in the IT equipment is likely to support the low 100 V AC, high 240 V AC, single-phase sources of 120 V and 240 V, and three-phase sources with voltages of 120 V, 208 V and 240 V. By standards set forth by the Server System Infrastructure Forum [7], a PSU rated for 120–127 V should operate normally at voltages ranging from 90 to 140 V. A PSU rated for 200–240 V should operate normally on input voltage from 180 to 264 V. Real design margins are even somewhat broader because of the need to handle input voltages from any country around the world. The power output from the PSU may even be automatically limited by input voltage to protect the PSU and internal circuitry from damage if connected to the lower voltage range.

2.3.2 Input Frequency within Allowable Ranges

Power supplies for IT equipment are typically designed for universal operations. That means a typical PSU can operate normally at frequencies from 47 to 63 Hz to accept utility power at both 50 Hz and 60 Hz. Moreover, the UPS should be able to regulate output frequency to meet the PSUs specification range of 47–63 Hz for all frequency variations in the AC power source.

2.3.3 Sufficient Input Power to Compensate for Power Factor

Power factor is the ratio of real power (W) to apparent power (VA), where real power is the capacity of the circuit to perform work, and apparent power is the product of the current and voltage of the circuit, which also includes current affected by reactive compounds. A power factor of 1 indicates that the voltage and current peak together, which means that the VA and watt values are the same. Poor power factor had been known to cause failed neutral conductors, overheated transformers and, in the worst cases, building fires. The power factor of a circuit is influenced by the type of equipment being powered:

- Circuits containing only resistive elements have a power factor of 1 (Figure 2.5).
- Circuits containing inductive elements have what is known as a "lagging" power factor because the current waveform lags the voltage waveform

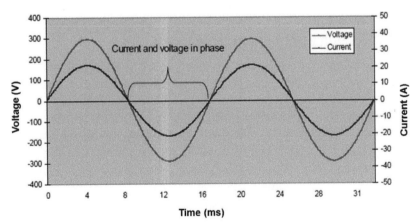

Figure 2.5   Resistive loads [7].

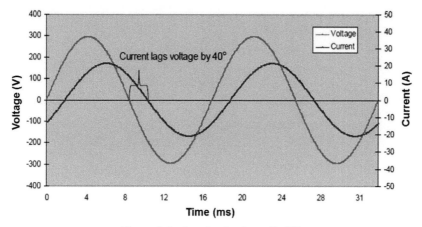

Figure 2.6   Lagging loads profile [7].

(Figure 2.6). Inductive elements consume reactive power. A consumption of reactive power by a load or generator can decrease, depending on the ratings of the load or generator, the voltage at its connection point with the network. If the voltage is reduced, the currents flowing through the generator/load have to be increased to maintain the same power generation/demand. High electrical currents imply high power losses.

- Circuits containing many capacitive elements have a "leading" power factor. The voltage waveform lags the current waveform. Capacitive elements generate reactive power. Generation of reactive power can increase the voltage at the connection point of an electrical load or generator. Conventionally, capacitive reactive power flows throughout the network is not permitted for voltage stability issues.
- Circuits containing a mixture of reactive components typically have what is known as a distortion power factor. The current drawn on the input is a mixture of the fundamental frequency, as well as several harmonic frequencies (Figure 2.7).

Power supplies used in IT equipment generally fit into this last category (harmonic load, Figure 2.7) due to the variety of power sources, load power profiles and power conversion systems based on electronic power devices. However, the current drawn from the source is much less distorted than that shown before. The harmonic distortion and the power factor are directly related. Thus, the higher the power factor of a device is, the lower the harmonic distortion.

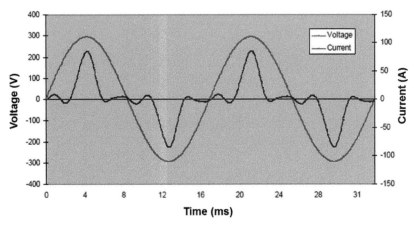

Figure 2.7　Harmonic load profile [7].

Nowadays, the PSUs used in IT equipment have a power factor trending towards unity, because of the need to minimize reactive currents for cost reduction in filters and cables, among others. The size of these components is directly related to the magnitude of the currents flowing through them, and therefore to the magnitude of reactive currents due to the presence of inductive or capacitive consumptions/generations, respectively, affecting the power factor of the installation. Moreover, and as previously noted, harmonic current content in the AC source feeding the IT loads is related to the power factor of the installation. Therefore, a power factor of 0.9 would be considered acceptable, 0.95 would be typical, and a value of 0.99 would be excellent. Since the power factor in a typical Data Centre is still less than unity, it is necessary to supply slightly more apparent power to get the real power needed.

2.3.4 Transfer to Backup Power Faster than PSU "Hold-up" Time

According to IT equipment standards, minimum hold-up time at fully rated output power is one cycle. At 50 Hz, this translates into 20 ms, while at 60 Hz, it would be 16.7 ms. Since most IT equipment is designed for the global market, the minimum hold-up time is 20 ms and may be longer at lighter loads. A related issue with respect to hold-up time is the peak inrush current (Table 2.3) required to charge up the capacitor that provides the ride-through capability. When first connected to an AC power source, the equipment temporarily draws a large inrush current that can last for 2–10 ms and be as much as 10–60 times the normal operating current.

Table 2.3 Example IT equipment nominal and peak inrush current

Equipment	Maximum Nominal Current	Peak Inrush Current
HP Proliant DL 360 G4 – IU server	2.4 A	61 A for 3 ms
HP Proliant e-class blade server	1.6 A	100 A for 2 ms
IBM BladeCenter, fully loaded	23.7 A	200 A for 4 ms
IBM x-series 260	4.9 A	120 A for 4 ms
Cisco 3825 Router	2.0 A	50 A for 10 ms

2.3.5 Protection from Damaging Power Conditions

PSUs are designed to handle voltage that sags 10% below nominal specification or surges 10% above, without loss of function or performance. If the nominal voltage range is 200–240 V, the PSU will operate normally when input voltage is as low as 180 V or as high as 264 V. The PSU is also required to handle surges of 30% from the midpoint of nominal for 0.5 cycles (8–10 ms). These tolerances are well defined by the Information Technology Industry Council curve shown in Figure 2.8 [8]. Voltage conditions within the upper and lower boundaries are safe. Below this zone is a low-voltage area where the

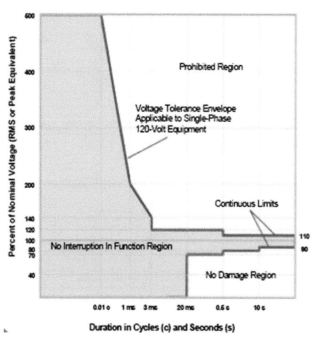

Figure 2.8 Range of power conditions that IT equipment can tolerate [8].

PSU would not be expected to operate normally, but it would not be harmed either. Above is the prohibited region, where voltage conditions could damage the equipment.

Moreover, IEEE Standards [9, 10] define and illustrate power quality disturbances and how to prevent them. Thus, Figure 2.9 summarizes the power disturbances and provides possible solutions to mitigate the effects that these problems can have on Data Centre operations.

Disturbance category	Wave form	Effects	Possible causes	Possible solutions
1. Transient				
Impulsive		Loss of data, possible damage, system halts	Lightning, ESD, switching impulses, utility fault clearing	TVSS, maintain humidity between 35 – 50%
Oscillatory		Loss of data, possible damage	Switching of inductive/capacitive loads	TVSS, UPS, reactors/ chokes, zero crossing switch
2. Interruptions				
Interruption		Loss of data possible, damage shutdown	Switching, utility faults, circuit breaker tripping, component failures	UPS
3. Sag / undervoltage				
Sag		System halts, loss of data, shutdown	Startup loads, faults	Power conditioner, UPS
Undervoltage		System halts, loss of data, shutdown	Utility faults, load changes	Power conditioner, UPS
4. Swell / overvoltage				
Swell		Nuisance tripping, equipment damage/reduced life	Load changes, utility faults	Power conditioner, UPS, ferroresonant "control" transformers
Overvoltage		Equipment damage/reduced life	Load changes, utility faults	Power conditioner, UPS, ferroresonant "control" transformers
5. Waveform distortion				
DC offset		Transformers heated, ground fault current, nuisance tripping	Faulty rectifiers, power supplies	Troubleshoot and replace defective equipment
Harmonics		Transformers heated, system halts	Electronic loads (non-linear loads)	Reconfigure distribution, install k-factor transformers, use PFC power supplies
Interharmonics		Light flicker, heating, communication interference	Control signals, faulty equipment, cycloconverters, frequency converters, induction motors, arcing devices	Power conditioner, filters, UPS
Notching		System halts, data loss	Variable speed drives, arc welders, light dimmers	Reconfigure distribution, relocate sensitive loads, install filters, UPS
Noise		System halts, data loss	Transmitters (radio), faulty equipment, ineffective grounding, proximity to EMI/RFI source	Remove transmitters, reconfigure grounding, moving away from EMI/RFI source, increase shielding filters, isolation transformer
Voltage fluctuations		System halts, data loss	Transmitters (radio), faulty equipment, ineffective grounding, proximity to EMI/RFI source	Reconfigure distribution, relocate sensitive loads, power conditioner, UPS
Power frequency variations		System halts, light flicker	Intermittent operation of load equipment	Reconfigure distribution, relocate sensitive loads, power conditioner, UPS

Figure 2.9   Summary of disturbances with solutions [11].

References

[1] RenewIT, "RenewIT tool," 2016. [Online]. Available: www.renewit-tool.eu.
[2] Oró, E. (2014). Deliverable D2.1 Requirements for IT equipment in Data Centre operation.
[3] "Thermal guidelines for Data Processing environments," ASHRAE Datacom Series 1 – Third Edition.
[4] "Intel," [Online]. Available: https://www-ssl.intel.com/content/dam/www/public/us/en/documents/datasheets/xeon-e5-v2-datasheet-vol-1.pdf. [Accessed 2 December 2013].
[5] "Intel," [Online]. Available: https://www-ssl.intel.com/content/dam/www/public/us/en/documents/datasheets/xeon-e5-v2-datasheet-vol-1.pdf. [Accessed 2 December 2013].
[6] ASHRAE, "Gaseous and particle contamination guidelines for data centers," 2011.
[7] Mitchell-Jackson, J., Koomey, J., Nordman, B., and Blazek, M. (2003). Data center power requirements: measurements from Silicon Valley. *Energy*, 28, 837–850.
[8] I. T. Committee, "Information Technology Industry Council".
[9] I. Std, "IEEE Recommended Practice for Monitoring Electric Power Quality".
[10] I. Std, "IEEE Recommended Practice for Powering and Grounding Sensitive Electronic Equipment".
[11] Syemour, J. "The seven types of power problems," Schneider Electric. White Paper 18.

3

Environmental and Economic Metrics for Data Centres

Jaume Salom and Albert Garcia

Catalonia Institute for Energy Research – IREC, Spain

3.1 About Metrics in Data Centres

With escalating demand and rising energy prices, it is essential for the owners and operators of these mission-critical facilities to assess and improve Data Centre performance. Metrics are an important way to understand the economic and environmental impact of Data Centres and the results of applying energy efficiency measures or integrating renewable energy sources (RES). A wide range of energy-efficiency metrics and key performance indicators (KPIs) has been developed over the past years. In this framework, eight European projects with the common topic of energy efficiency in Data Centres have formed the Smart City Cluster Collaboration with the objective to define and to agree common metrics and methodologies. An official liaison has been established between the Smart City Cluster Collaboration with the join technical committee ISO/EIC JTC1-SC39 on "Sustainability for and by Information Technology" to collaborate in metrics standardisation activities.

Low-level measurement of operational variables such as temperature and relative humidity in Data Centre IT rooms are key aspects as they are part of the service-level agreements (SLA) in IT services. Several metrics addressing the issue if temperature and relative humidity are within the allowed boundaries are extensively described in the literature [38] and reviewed in the context of the [40]. Being aware of its relevancy, the focus of this chapter refers on metrics at facility level.

A diversity of views exits [13] about the convenience of using global synthetic indicators defined as a weighted combination of several basic metrics.

41

Other views prefer to show each metric separately supported with some graphic format. For example, a spider chart is able to provide a combined visualization of several metrics in one graph. The number of metrics (axes) may vary depending on the selection of metrics chosen by the Data Centre operator.

Figure 3.1 shows an example of a spider chart with 5 axes. Once the metrics are selected, it is also needed identifying a start and end point for each axis. In some cases, there are theoretical maximum and minimum values (e.g. share of renewables can only range from 0 to 1). In other cases, there are not clear maximum values (e.g. there is no maximum for PUE). Therefore, the axis ends will have to be established based on target values or other estimates. A spider chart using three axes is adopted by Future Facilities to define the ACE Performance Score [17], shown in Figure 3.2. The ACE Performance Score triangle compares the difference between the actual Data Centre's performance

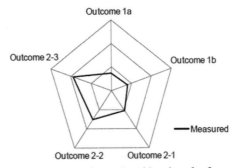

Figure 3.1 Example of generic spider chart for five metrics.

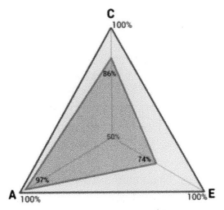

Figure 3.2 ACE Performance Score [17].

with the expected output when it was designed in three aspects: availability, capacity and efficiency.

Other companies are proposing different forms of dashboard to report key efficiency metrics for their companies. For example, Ebay implemented Digital Service Efficiency (DSE) methodology to see the full cost, performance and environmental impact of customer buy and sell transactions, giving eBay a holistic way to balance and tune its technical infrastructure [10]. A dashboard displays the Ebay's specific DSE measures over a set of user-selected time periods (see Figure 3.3).

A holistic framework helps the operator keep in mind the effects on all metrics simultaneously and is a way to grasp multiply metrics for Data Centre collectively [19]. Other important aspects that can influence energy metrics such as availability, capacity, economic costs or security issues in Data Centres can be also considered in a holistic perspective.

To assess the environmental impact of the Data Centres among other human-promoted activities is a key issue to assure the sustainability of our society and to prevent global warming effects. In the point of view of stakeholders responsible to take decisions about actions to carry out in Data Centres, and particularly related with implementation of measures to improve the energy efficiency and share of renewables, the economic feasibility must be included as one of the main indicators, including the costs of carbon emissions, if it is considered. This chapter proposes a methodology for cost-environmental analysis centred in energy use of Data Centres, aiming

Figure 3.3 Digital Service Efficiency (DSE) dashboard from [10].

to analyse both the economic and environmental impact of any efficiency measure. This will allow to optimize and to select the optimal solution among several options. This comparative framework needs to be based on relevant metrics, both environmental and financial, which are described in Section 3.3 of this chapter. Moreover, other relevant energy-efficiency metrics or KPI's to characterise the use of renewables in Data Centres are introduced as well.

3.2 Data Centre Boundaries for Metrics Calculation

3.2.1 Definition of Boundaries

In a future perspective of energy-producing premises, relying on renewable energy sources or high-efficient systems, the Data Centre infrastructure can be categorized with different system blocks, as shown in Figure 3.4.

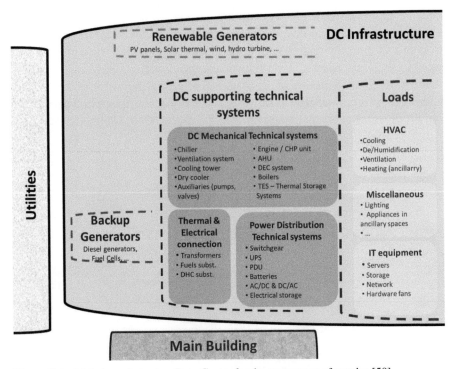

Figure 3.4　Main boundaries in a Data Centre for the assessment of metrics [50].
Source: IREC.

- Data Centreloads
  - IT workload (IT equipment)
  - Miscellaneous loads (i.e.lighting)
  - Mechanical or HVAC load (i.e. refrigeration for white space and ancillary spaces)
- Data Centre supporting technical systems
  - Power distribution systems. Systems which distribute power to IT equipment and other elements in the Data Centre.
  - Mechanical systems. Components as chillers, CRAC, air handling units, pumps, cooling towers, etc., aiming to cover HVAC needs.
  - Technical connection subsystems. Systems for connecting with electrical and thermal utilities as for example the power transformers to connect to the main grid or the heat exchangers and additional hydraulic elements to connect with a district cooling network.
- On-site renewable energy generators, which are renewable energy systems placed in the Data Centre footprint, as for example photovoltaic (PV) panels or micro-wind turbines.
- Backup system. Component to supply power in case that the electricity grid presents a problem, as diesel generators, batteries, fuelcells or flywheels.

3.2.2 Energy Flows

In a so-called Net Zero Energy Data Centre, highly efficient and renewable sources driven on-site energy supply systems will be adopted, such as cogeneration units, PV panels or solar thermal collectors. Thereby, a wider group of technologies is interposed between the local utilities networks from which the driven energy carriers are imported, like gas for a cogeneration unit, and the equipment which receive the energy from the supply systems before feeding it into the IT equipment. Such technologies can support the Data Centre operation by supplying electricity but also heat for cooling purposes, like in the case of tri-generation or solar cooling systems.

Some of the energy supply systems or heat recovering systems from the IT equipment, all of them being part of the Data Centre infrastructure, can export energy to the utility or can be shared with the main building where the Data Centre is installed. Figure 3.5 shows a simplified scheme where backup generators have been eliminated from the scheme and energy flows have been divided in different energy carriers. The scheme is simplified eliminating

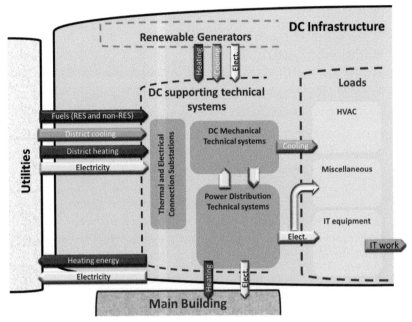

Figure 3.5   Simplified scheme of energy flows in a Data Centre infrastructure with electricity and cooling energy needs [50]. *Source:* IREC.

the energy flows which are less relevant for a Data Centre. Only electricity and cooling needs are considered, although in some Data Centre infrastructures, some heating loads could be required for ancillary spaces. Regarding exported energy, Data Centre exports of cooling to the utilities or other buildings is not considered, as it is an intensive cooling consumer.

Delivered and exported energies have to be calculated separately for each energy carrier. Delivered energy from outside the Data Centre infrastructure can be in the form of electricity, heating or cooling from a district network or fuels, both renewable and non-renewable. On-site renewable energy generators without fuels mean the electric and thermal energy (heating or cooling) produced by solar collectors, PV, wind or hydro turbines. The thermal energy extracted from ambient or other renewable environments (e.g. sea water) through heat exchangers is also considered on-site renewable energy. Renewable fuels, as for example biomass or biogas, are not included in on-site renewables, but they are taken into account as renewable part of the delivered energy, in form of renewable fuel. Figure 3.6 depicts the nomenclature of the energy flows used in this book in a simplified way.

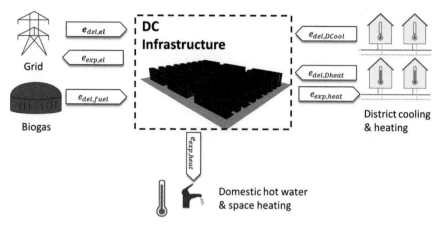

Figure 3.6 Simplified scheme of energy flows in a Data Centre infrastructure with nomenclature.

3.3 Metrics for Cost-Environmental Analysis

3.3.1 Environmental Impact Metrics

3.3.1.1 Data Centre primary energy

The main metric to evaluate the energy consumption in the Data Centre considering environmental issues is the Data Centre primary energy. The Data Centre primary energy accounts for the energy that has not been subjected to any conversion of transformation process which also receives the name of source energy, as for example in [19].

The primary energy indicator sums up all delivered and exported energy, for all the energy carriers, into a single indicator with corresponding (national, regional or local) primary energy weighting factors. By default, the non-renewable Data Centre primary energy is used. For a given energy carrier, the non-renewable primary energy factor is the non-renewable primary energy divided by delivered energy. The non-renewable primary energy is defined as the energy required to supply one unit of delivered energy, taking into account of the non-renewable energy required for extraction, processing, storage, transport, generation, transformation, transmission, distribution and any other operations necessary for delivery to the Data Centre in which the delivered energy will be used. Notice that the non-renewable primary energy factor can be less than unity if renewable energy has been used.

$$PE_{\text{DC,nren}} = \int Pe_{\text{DC,nren}}(t) \cdot \mathrm{d}t \qquad (3.1)$$

$$Pe_{\text{DC,nren}}(t) = \sum_i (e_{\text{del},i}(t) \cdot w_{\text{del,nren},i}(t)) - \sum_i (e_{\text{exp},i}(t) \cdot w_{\text{exp,nren},i}(t))$$
(3.2)

If we consider the energy flows which are common in a Data Centre according to Figure 3.6 (excluding cooling as exported energy option), primary energy can be calculated using Equation (3.3)[1].

$$Pe_{\text{DC,nren}} = [(e_{\text{del,el}} \cdot w_{\text{del,nren,el}}) + (e_{\text{del,fuel}} \cdot w_{\text{del,nren,fuel}}) +$$
$$+ (e_{\text{del,DHeat}} \cdot w_{\text{del,nren,DHeat}}) + (e_{\text{del,DCool}} \cdot w_{\text{del,nren,DCool}})] -$$
$$- [(e_{\text{exp,el}} \cdot w_{\text{exp,nren,el}}) + (e_{\text{exp,heat}} \cdot w_{\text{exp,nren,heat}})]$$
(3.3)

According to this definition, Net Zero Energy Data Centre has an exact performance level of 0 kW·$h_{\text{PE,nren}}$ non-renewable primary energy. A plus energy or positive energy Data Centre will have a negative value (<0) of non-renewable primary energy.

In some cases, instead of using the non-renewable energy factors, the total energy weighting factors may be used. In this case, the total Data Centre Primary Energy is the one that accounts both for the non-renewable and renewable primary energy. The total primary energy factors always are equal or exceed 1.0.

$$PE_{\text{DC,tot}} = \int Pe_{\text{DC,tot}}(t) \cdot dt$$
(3.4)

$$Pe_{\text{DC,tot}}(t) = \sum_i (e_{\text{del},i}(t) \cdot w_{\text{del,tot},i}(t)) - \sum_i (e_{\text{exp},i}(t) \cdot w_{\text{exp,tot},i}(t))$$
(3.5)

Remarks on Weighting Factors

Quantification of proper conversion factors is not an easy task, especially for electricity and thermal networks as it depends on several considerations, e.g., the mix of energy sources within certain geographical boundaries (international, national, regional or local), average or marginal production, present or expected future values and so on. In general, there are no correct conversion factors in absolute terms. Rather, different conversion factors are possible, depending on the scope and the assumptions of the analysis. This leads to the fact that "strategic corrected" weighting factors may be adopted in order to find a compromise agreement.

---

[1]Notice that matemathical formulation has been simplified. All the terms are time dependent.

Furthermore, "strategic factors" may be used in order to include considerations not directly connected with the conversion of primary sources into energy carriers. Strategic factors can be used to promote or discourage the adoption of certain technologies and energy carriers, as it has been proven in [29] for the case of Net Zero Energy Buildings.

Generally speaking, weighting factors can be generally time dependent, as the share of renewables is dependent on the season and the period of the day. In [18], the seasonal and daily variation of conversion factors for the electricity mix in several European countries has been analysed. However, they are subjected to changes due to the planned increase in the share of renewables towards 2050. Usually, mean annual national and regional factors are available subjected to different regional or country approaches [35]. In case of absence of national/regional factors, European or global factors can be used as reference. The FprEN ISO 52000-1:2016 [8] proposes default values for primary energy weighting factors which can be used for reference. In [19], global average factors are provided and its use is recommended for comparison across different regions in the world.

Remarks on Exported Energy Accounting

One key aspect that affects the computation of the defined Data Centre primary energy metrics is the methodology for accounting the exported energy which is the delivered energy by technical Data Centre systems through the boundary and used outside the Data Centre boundary. Generally speaking, the exported energy can be both in forms of electricity or thermal energy (heating and cooling). It accounts for the energy generated on-site which do not match instantaneously the energy needs and therefore needs to be exported. By default, the weighting factors for the exported energy for energy carrier are equal to the factor for the delivered energy, if not specifically defined in other way. However, the factors to apply to the exported energy are not completely clear defined. FprEN ISO 52000-1:2016 proposes different factors for the electricity depending if it is exported to the grid, exported for the immediate use or exported temporary to be reused later, but no proposal for heat or cooling exported flows is presented. In a Data Centre, excess heat could be from a cogeneration system or from the ability to re-use heat from the IT white space. Recommendation by the German Heat and Power Association (AGFW) [3] is to consider $w_{exp,nren,heat} = 0.0$ because the Data Centre waste heat is a by-product of a process which is not related to the power supply industry. In calculations in Chapter 7, by default $w_{exp,nren,heat}$ has a value different from zero as considered the amount of heat that can be effectively used by a third entity in the primary energy balance.

3.3.1.2 Data Centre CO_2 emissions

Data Centre CO_2 emissions can be computed using adequate conversion factors for each energy carrier, using similar methodology that used to compute Data Centre primary energy. The weighting factors to be used are the CO_2 emission coefficient which is the quantity of CO_2 emitted to the atmosphere per unit of energy, for a given energy carrier. The CO_2 emission weighting factor can also include the equivalent emissions of other greenhouse gases. See [18] about CO_2 emission coefficients from the electricity grid.

$$EM_{DC,CO_2} = \int em_{DC,CO_2}(t) \cdot \mathrm{dt} \tag{3.6}$$

$$em_{DC,CO_2}(t) = \sum_i (e_{del,i}(t) \cdot w_{del,CO_2,i}(t)) - \sum_i (e_{exp,i}(t) \cdot w_{exp,CO_2,i}(t)) \tag{3.7}$$

3.3.1.3 Data Centre water consumption

Depending on the design of the cooling system of Data Centres, they can require significant amounts of water which have to be accounted for, especially if Data Centre is located in a site with limited availability of water. Then, the water consumption is a valued that needs to be considered which can be measured continuously or integrated over a period of time using Equation (3.8) ($Water_{DC}$).

$$Water_{DC} = \int water_{DC}(t) \cdot \mathrm{dt} \tag{3.8}$$

3.3.2 Financial Metrics

3.3.2.1 Methodological reference framework

A methodology framework is proposed which gives the basis to consider calculation of cost-optimal levels for both financial and macroeconomic viewpoint. The methodological framework is adapted from the existing *EN-15459* [7], together with [12, 14]. In the financial calculation, the relevant prices to take into account are the prices paid by the end-user

(Data Centre owner or operator) including all applicable taxes and charges. For the calculation at macroeconomic level, an additional cost category taken into account the costs of greenhouse gas emissions is introduced and applicable charges and taxes should be excluded.

3.3.2.2 Global cost

The total cost of ownership (TCO) (or the total global cost) is defined as the present value of the initial investment costs (CAPEX), sum of running costs or operational expenditures (OPEX), and replacement costs (referred to the starting year), as well as residual values, if applicable. Thus, it can be understood as a way to quantify the financial impact of any capital investment regarding IT business (see Figure 3.7). The TCO is used to assess the true total costs of building, owning and operating a Data Centre physical facilities [45]. The basic principle for the calculation of the global cost is made for the system – or each component j of the system – considering the initial investment C_I, the present value of annual costs for any year i

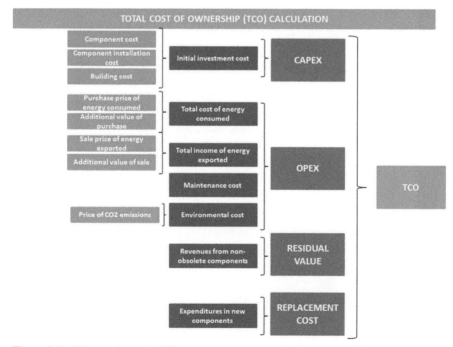

Figure 3.7 Scheme about the different parts composing the Total Cost of Ownership (TCO).

and the final value of any component or system. General expression for the calculation of the global cost for a period τ is Equation (3.9).

$$\text{TCO}\,(\tau) = C_I + \sum_j \left[\sum_{i=1}^{\tau} (C_{a,i}\,(j) \cdot R_d(i)) - V_{f,\tau}(j) \right] \qquad (3.9)$$

where

> C_I are the initial investment costs or CAPEX
> $C_{a,i}\,(j)$ are the annual costs for the year i and component j, including operational and replacement costs, namely OPEX
> $R_d(i)$ is the discount rate for the year i
> $V_{f,\tau}(j)$ is the final value of component j, if any, at the end of the period.

3.3.2.3 CAPEX: capital expenditure

This is the amount of money used to acquire assets or improve the useful life of existing assets. In general terms, CAPEX would include server purchasing costs, construction costs of a new Data Centre and any investment realized to improve or enlarge the Data Centre facility. Market survey developed by [2] in 2012 estimates the investment costs for a traditional Data Centre and for modular Data Centres. The average values in € per unit of IT installed capacity power are 10 784 €/kW and 6 470 €/kW for traditional and modular Data Centres, respectively.

$$\text{CAPEX} = \sum_{j=1}^{NC} [\text{CC}_j + \text{CI}_j] + \text{CCDC} \qquad (3.10)$$

where

> NC are the number of components defining the energy facility
> CC_j is the investment cost of component j
> CI_j is the installation cost of component j
> CCDC is the cost of the main building the Data Centre, mainly building cost.

3.3.2.4 OPEX: operating expenditure

Operating expenditure (OPEX) is the ongoing cost for running a product, business or system. In the case of Data Centres, OPEX (see Equation (3.11)) includes [4] the following:

- Energy costs: electricity costs and other energy carrier costs
- License costs
- Maintenance costs
- Labour costs
- Utilities costs, except energy cost
- Replacement cost

Energy costs should be calculated for each energy carrier and prices can be temporary dependent. The costs of energy consumed for a given period of time can be calculated with Equation (3.12) being the weighting factors the economic cost per unit of delivered energy or the income per unit of exported energy.

$$\text{OPEX} = \text{OPEX}_{\text{EC}} + \text{OPEX}_{\text{CM}} + \text{OPEX}_{\text{REC}} + \text{OPEX}_{\text{CO}_2} \quad (3.11)$$

$$\text{OPEX}_{\text{EC}} = \int C e_{\text{DC}}(t) \cdot dt = \int \sum_i (e_{\text{del},i} \cdot w_{\text{del},\text{€},i})$$

$$- \sum_i (e_{\text{exp},i} \cdot w_{\text{exp},\text{€},i}) \quad (3.12)$$

$$Ce_{\text{DC}}(t) = \sum_i (e_{\text{del},i}(t) \cdot w_{\text{del},\text{€},i}(t)) - \sum_i (e_{\text{exp},i}(t) \cdot w_{\text{exp},\text{€},i}(t))$$

One should consider as negative expenses, additional economic incomes derived, for example, from energy management actions or better management of available capacity.

3.3.3 Cost-Efficiency Analysis

Cost-efficiency analysis can be done by establishing a comparative methodology framework. The methodology specifies how to compare energy-efficiency measures, measures incorporating renewable energy sources and packages of such measures in relation to their environmental performance and the cost attributed to their implementation.

In order to have an appropriate evaluation among a set of solutions, one could opt for a graphical representation. The x-axis represents the Data Centre primary energy or other environmental index (CO_2 emissions or water consumption) and the y-axis the TCO. Each point represents the values (TCO and $PE_{\text{DC,nren}}$) that result when a combination of compatible energy efficiency and energy supply measures is applied to a Data Centre.

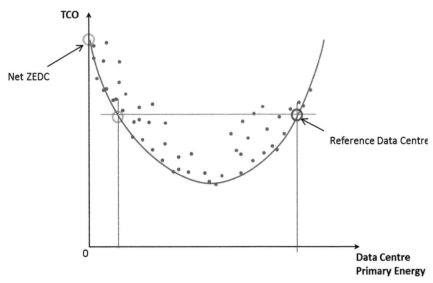

Figure 3.8   Graphical representation for cost-optimal analysis. *Source:* IREC.

If cost-effective measure is applied to a reference Data Centre and primary energy is used as environmental index (each point in Figure 3.8 represents a solution), it is possible to identify the following:

- The cost-optimal configuration. The solution having the minimum global cost.
- A set of solution which are cost effective. The ones having less primary energy and equal or less TCO than the reference case.
- The additional global cost needed to reach a Net Zero Energy Data Centre.

3.4 Energy Efficiency and Renewable Energy Metrics

3.4.1 Power Usage Effectiveness (PUE)

Power usage effectiveness, known as PUE, is the most popular and well-known key performance indicator used in the Data Centre industry which aims to quantify the efficient use of energy in the form of electricity. PUE is defined as ratio of the Data Centre total energy consumption to information technology equipment energy consumption, calculated, measured or assessed across the same period. The reader is referred to the standard ISO/IEC 30134-2:2016 Information Technology – Data Centres – Key Performance Indicators – Part 2:

Power usage effectiveness (PUE) to know more details about the categories of PUE and the instructions for its measurement. Due to the heterogeneity of Data Centre facilities, it is not recommended that PUE values from different Data Centres are compared directly. PUE should principally be used to assess trends in an individual facility over time and to determine the effects of different design and operational decisions within a specific facility [23]. In the framework of this book, advanced resolution of PUE (Category 3 – PUE_3) is used. PUE_3 is characterized by the measurement of the IT load at the IT equipment in the Data Centre.

As a result that several energy carriers can be used to determine the total energy consumed for the Data Centre operation, here PUE is calculated as the total primary energy consumption of the Data Centre ($PE_{DC,tot}$) divided by the total primary energy delivered to the IT equipment ($PE_{DC,tot,IT}$). PUE is a metric to identify how efficient the electricity is used from the Data Centre control to the IT equipment. Therefore, it gives the relation of the extra amount of energy consumed in order to keep the servers working properly. PUE is defined in coherence with the standard [23] where the energy that is reused is not subtracted from the total.

$$\text{PUE} = \frac{PE_{DC,tot}}{PE_{DC,tot,IT}} = \frac{\int Pe_{DC,tot}(t) \cdot dt}{\int Pe_{DC,tot,IT}(t) \cdot dt}$$

where

$$Pe_{DC,tot}(t) = \sum_i e_{del,i}(t) \cdot w_{del,tot,i}(t) - \sum_i e_{exp,el}(t) \cdot w_{exp,tot,el}(t) +$$
$$+ e_{ren,i}(t) \cdot w_{del,tot,i}(t)$$

$$Pe_{DC,tot,IT}(t) = e_{del,IT}(t) \cdot w_{tot,IT}(t)$$

$$w_{tot,IT} = \frac{\text{Primary energy to produce electricity} + \text{Purchased electricity}}{\text{All electricity at site}}$$

$$w_{tot,el} = \frac{e_{del,i} \cdot w_{del,tot,i} + e_{grid} \cdot w_{del,tot,el}}{e_{del,i} + e_{grid}}$$

3.4.2 Renewable Energy Ratio

The renewable energy ratio (RER) is the metric that allows calculating the share of renewable energy use in a Data Centre. The renewable energy ratio is calculated relative to all energy use in the Data Centre, in terms of total primary energy and it is analogous to the Total PE Percent described in [39] and [31].

$$\mathrm{RER_{EP}} = \frac{\sum_i e_{\mathrm{ren,i}} + \sum_i [(w_{\mathrm{del,tot,i}} - w_{\mathrm{del,nren,i}}) \cdot e_{\mathrm{del,i}}]}{\sum_i e_{\mathrm{ren,i}} + \sum_i (w_{\mathrm{del,tot,i}} \cdot e_{\mathrm{del,i}}) - \sum_i (w_{\mathrm{exp,tot,i}} \cdot e_{\mathrm{exp,i}})}$$

$$(3.13)$$

The calculation considers that exported energy compensates delivered energy. By default, it is considered that exported energy weighting factors compensate the electrical grid mix or the district heating or cooling network mix, if any, in the case of thermal energy. Regarding consideration of exported energy in the computation, the reader is referred to the remarks in Section 3.1.1.

For the calculation of the RER, all renewable energy sources have to be accounted for. These include solar thermal, solar electricity, wind and hydroelectricity, renewable energy captured from ambient heat sources by heat pumps and free cooling, renewable fuels and off-site renewable energy. For on-site renewable energy, the total primary energy is 1.0. The amount of energy captured by heat pumps from ambient heat sources should be according EU Directive 2009/28/EC and Commission Decision 2013/114/UE [11]. Figure 3.9 depicts the boundaries for the energy use system and the main flows that need to be considered to compute RER.

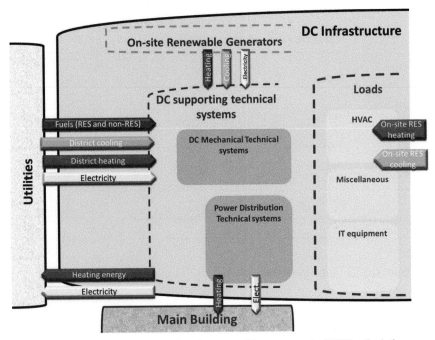

Figure 3.9   Use system boundary for renewable energy ratio (RER) calculation.

3.4.3 Renewable Energy Factor

In contrast to renewable energy ratio (RER) as defined in previous section, other KPI is commonly used to characterise the percentage of renewable energy over total Data Centre energy. The reader is referred to the standard ISO/IEC 30134-2:2016 Information Technology – Data Centres – Key Performance Indicators – Part 3: Renewable energy factor (REF) to know more about the computation of the REF. Main difference between the RER is that REF shall have a maximum value of 1.0, as on-site generation of renewable energy beyond the need of the Data Centre should not be accounted for REF calculations. In addition, only the renewable energy that is owned and controlled by the Data Centre is accounted, meaning that the energy for which the Data Centre owns the legal right, as it is certified by providers or renewable energy certificates. On-site generation which certificates are sold together with the generated energy must not be taken into account for the REF calculation. Using the same nomenclature, REF can be formulated as follows:

$$\mathrm{REF_{EP}} = \frac{\sum_i e_{\mathrm{ren,i}} + \sum_i [(w_{\mathrm{del,tot,i}} - w_{\mathrm{del,nren,i}}) \cdot e_{\mathrm{del,i}}]}{\sum_i e_{\mathrm{ren,i}} + \sum_i (w_{\mathrm{del,tot,i}} \cdot e_{\mathrm{del,i}})} \qquad (3.14)$$

3.5 Capacity Metrics

3.5.1 Introduction

Data Centres are planned and designed to meet the maximum future estimated capacity for the IT requirements in a period of time (up to 10 years, usually). This means that at day one, the investments usually donot fulfil the ultimately design capacity and this is expected to increase from the start-up load to the expected final value. The unused capacity in IT and power infrastructures represents avoidable capital costs and operational costs, including maintenance and energy costs, too. According to several sources, the degree of utilisation of Data Centres is between 50% [30] and 70% [17]. Many operators have the standard practice of utilizing only a fraction, such as 80% or 90% of the installed capacity [30], assuming that operating the system at less than full power will maximize overall reliability. The traditional model for planning Data Centres infrastructures is illustrated in Figure 3.10 which depicts the waste of resources to oversizing. Deploying an adaptable physical infrastructure, the waste due to oversizing can be reduced, as it is shown graphically in Figure 3.11.

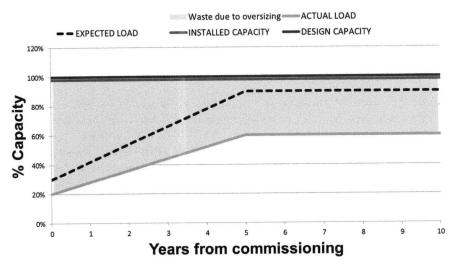

Figure 3.10 Design capacity and expected load in a planning model with all the capacity available since day one. Adapted from [30].

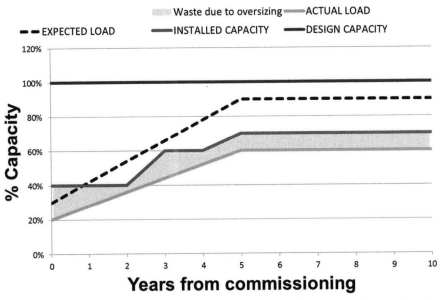

Figure 3.11 Design capacity and expected load in a planning model with an adaptable physical infrastructure. Adapted from [30].

Then, having in mind that unused capacities have a big impact in the use of resources and in the economic profitability of a Data Centre, some metrics have been recently proposed to show the utilization of the available capacity and the capacity analysis starts to be introduced in DCIM. ACE score [17] shows the utilization of capacity but it is referred to the IT load Capacity. A capacity metric, named the Cooling Capacity Factor (CCF) is presented in [5]. The CCF is defined as the ratio of total running manufacturer's rated cooling capacity to 110% of the critical load. Ten percent is added to the critical load to estimate the additional heat load of lights, people, etc. The white paper [5] reveals average CCF of 3.9, which is very high compared to the ideal CCF = 1.2.

Therefore, capacity metrics intends to relate the actual peak IT power and the peak facility power with both the design and installed capacity, in case they are different. With these metrics, one can evaluate how to increase as much physical capacity in a Data Centre as possible (this is the capacity that owner–operator has already paid for). Failing to use this capacity will incur future capital and operational expenditure. In case of Data Centres incorporating renewable energy systems, the on-site generation system is also characterized by its capacity. Also, we can distinguish between the design capacity, meaning the ultimately generation capacity related with the design load, the actual installed capacity and the actual peak power of the generation system. Having in mind that generation power can be instantaneously higher than the load, this leads to the situation of exporting energy to the grid.

3.5.2 Capacity Metrics

The *connection capacity credit*, or *power reduction potential* [34], is defined as the percentage of grid connection capacity that could be saved compared to a reference case. Positive values of this index indicate a saving potential; negative values indicate a need to increase the grid connection capacity with respect to the reference case. It can be formulated in the Equation (3.15).

$$CC = 1 - (DR/DR_{ref}) \qquad (3.15)$$

Based on that concept, some capacity credit (CC) metrics are proposed for Data Centres, both focusing in the IT capacity and the total facility capacity. Notice that in the following proposed metrics, it is assumed that the actual facility peak power is greater than the generation peak power. If not, dimensioning rate (DR) should be used and some of the equations will need to be reformulated using DR.

IT Installed Capacity Credit

$$CC_{IT,inst} = 1 - (P_{IT}/P_{IT,inst}) \tag{3.16}$$

IT installed Versus Designed Capacity Credit

$$CC_{IT,des} = 1 - (P_{IT,inst}/P_{IT,des}). \tag{3.17}$$

Total Facility Installed Capacity Credit

$$CC_{fac,inst} = 1 - (P_{fac}/P_{fac,inst}) \tag{3.18}$$

Total facility Installed Versus Designed Capacity Credit

$$CC_{fac,des} = 1 - (P_{fac,inst}/P_{fac,des}) \tag{3.19}$$

where

$P_{IT,des}$ Designed IT power capacity
$P_{IT,inst}$ Installed IT power capacity
P_{IT} Actual IT peak power
$P_{fac,des}$ Designed total facility power capacity
$P_{fac,inst}$ Installed total facility power capacity
P_{fac} Actual total facility peak power

Space Capacity Credit

However, the physical capacity is dictated by the resource that is least available which could be space rather than cooling or IT power. Then, we can define an additional capacity metric related to the surface used, the so-called space capacity credit.

$$CC_{IT,m2} = 1 - \left(\frac{\text{Actual whitespace m}^2 \text{ occupied}}{\text{Design withespace m}^2} \right)$$

3.6 Examples

In this section, some illustrative examples have been provided to demonstrate the calculation of metrics for several Data Centre designs. Data Centre concepts presented in each example and the numerical values for each of the energy flows are based in rough annual calculations and estimations. For each of the examples, electrical and thermal schemes of the concept are presented together with a Sankey diagram to show the energy flows in the Data Centre. The primary energy factors and CO_2 emission factor used to calculate the metrics in the different examples are listed in Tables 3.1 and 3.2.

Table 3.1 Primary energy factors of the examples

		Primary Energy Factor (w)	Example 1	Example 2
Delivered	Non-Renewable	Electricity	2.135	2.135
		Biogas	–	–
		District Cooling	–	0.2
		On-site PV	0.0	–
		Aerothermal	0.0	0.0
	Total	Electricity	2.461	2.461
		Biogas	–	–
		District Cooling	–	1
		On-site PV	1.0	–
		Aero thermal	1.0	1.0
Exported	Non-Renewable	Electricity	2.135	2.135
		Heat	–	1.0
		District Cooling	–	0.2
	Total	Electricity	2.461	–
		Heat	–	1.0

Table 3.2 CO_2 emission factor of the examples

		CO_2 emission Factor (w)	Example 1	Example 2
Delivered $\frac{kg_{CO_2}}{kWh_{el}}$	CO_2	Electricity	0.348	0.348
		Biogas	–	–
		District Cooling	–	0
Exported $\frac{kg_{CO_2}}{kWh_{el}}$	CO_2	Electricity	0.348	0.348
		Biogas	–	–
		District Cooling	–	0

3.6.1 Example 1. PV System and Ice Storage

Example 1 corresponds to a concept where a vapour-compression chiller with wet cooling towers is used to produce cooling energy during summer. The electrical power required to drive the chiller can be purchased from the grid or generated by the photovoltaic system on the Data Centre footprint. A large ice storage tank is used for decoupling power and cooling generation from cooling demand (Figures 3.12 and 3.13).

Main Parameters of the System

The system uses a PV plant to satisfy approximately 80% of the energy consumption of the Data Centre. One part of the electricity generated by the PV panels is not self-consumed and therefore exported to the grid.

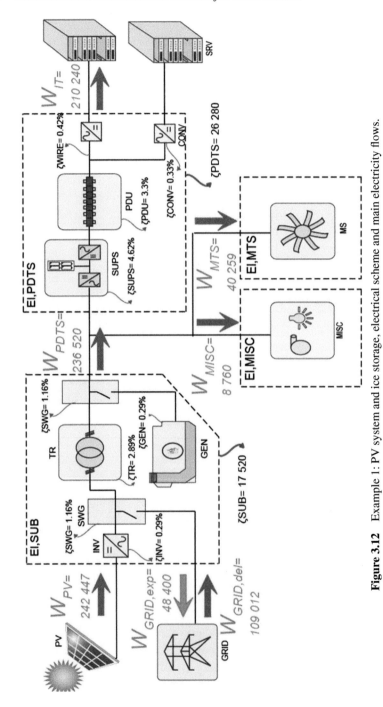

Figure 3.12   Example 1: PV system and ice storage, electrical scheme and main electricity flows.

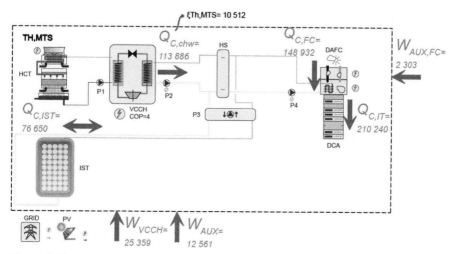

Figure 3.13 Example 1: PV system and ice storage, thermal scheme and main thermal and electricity flows linked to the mechanical technical systems.

The hypotheses for the facility power capacities are that the installed capacity is 20% higher than peak power value and the design capacity is the same as the installed capacity. As the example is located in Barcelona, it is assumed that the free cooling strategy is applied during 5374 hours. SEER values of 4.0 and 70.0 are assumed for the performance of the VCCH and the free cooling, respectively (Figure 3.14 and Table 3.3).

$$P_{IT} = 24\text{ kW} \quad P_{fac,G=0} = 41.2\text{ kW} \quad P_{PV} = 174\text{ kW}$$

$$P_{IT,inst} = 1.2 \cdot P_{IT} = 28.8\,\text{kW}_{el},\ P_{fac,inst} = 1.2 \cdot P_{fac} = 1.2 \cdot 144 = 172.8\,\text{kW}_{el}$$

$$P_{IT,des} = 1.1 \cdot P_{IT,inst} = 31.7\text{ kW}_{el},\ P_{fac,des} = 1.0 \cdot P_{fac,inst} = 172.8\text{ kW}_{el}$$

$$\text{SEER}_{VCCH} = 4.0 \quad \text{SEER}_{FC} = 70$$

$$W_{IT} = 210240\text{ kWh}$$

$$Q_{C,FC} = 161220\text{ kWh}_{th}$$

$$Q_{C,chw} = (Q_{C,IT} - Q_{C,FC}) + \xi_{th,MTS} = (210240 - 161220) + 52560$$

$$= 101580\text{ kWh}_{th}$$

$$W_{VCCH} = \frac{Q_{C,chw}}{\text{SEER}} = \frac{101580}{4.0} = 25395\text{ kWh}_{el}$$

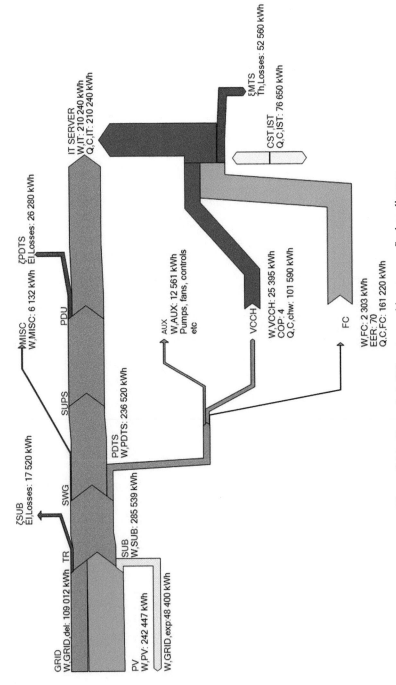

Figure 3.14   Example 1: PV system and ice storage. Sankey diagram.

Table 3.3 Example 1. Summary table of main energy flows

Symbol	Definition	kWh/year	PE nren	PE ren	PE tot
$W_{\text{GRID,del}}$	Delivered grid electricity	109 012	232 740	35 538	268 278
$W_{\text{GRID,exp}}$	Exported grid electricity	48 400	103 334	15 778	119 112
W_{PV}	On-site PV Electricity	242 447	0	242 447	242 447
$Q_{\text{C,FC}}$	Aero thermal energy, Free cooling	161 220	0	161 220	161 220

$$W_{\text{MTS}} = W_{\text{VCCH}} + W_{\text{AUX}} + W_{\text{AUX,FC}} = 25395 + 12561 + 2303$$
$$= 40259 \text{ kWh}$$
$$W_{\text{AUX}} = 12561 = 5131 + 7430 \text{ kWh (cooling tower +}$$
$$+ \text{ cooling distribution)}$$
$$W_{\text{GRID}} + W_{\text{PV}} = \zeta_{\text{SUB}} + W_{\text{PDTS}} + W_{\text{MISC}} + W_{\text{MTS}}$$
$$= 17520 + 236520 + 8760 + 40259 = 303059 \text{ kWh}$$
$$W_{\text{GRID}} = 60612 \text{ kWh} \quad W_{\text{PV}} = 242447 \text{ kWh}$$
$$W_{\text{GRID,del}} = 109012 \text{ kWh}$$
$$W_{\text{GRID,exp}} = 48400 \text{ kWh}$$

Non-Renewable Primary Energy

$$PE_{\text{DC,nren}} = E_{\text{del,el}} \cdot w_{\text{del,nren,el}} - E_{\text{exp,el}} \cdot w_{\text{del,nren,el}}$$
$$= (109012 - 48400) \cdot 2.135 = 60612 \cdot 2.135 = 129407 \text{ kWh}_{\text{PE,nren}}$$

Total Primary Energy

$$PE_{\text{DC,tot}} = E_{\text{del,el}} \cdot w_{\text{del,tot,el}} - E_{\text{exp,el}} \cdot w_{\text{exp,tot,el}} = 60612 \cdot 2.461$$
$$= 149166 \text{ kWh}_{\text{PE,tot}}$$

CO_2 emissions

$$EM_{\text{DC,CO}_2} = E_{\text{del,el}} \cdot w_{\text{del,CO}_2,\text{el}} - E_{\text{exp,el}} \cdot w_{\text{exp,CO}_2,\text{el}}$$
$$= 60612 \cdot 0.348 = 21093 \text{ kgCO}_2$$

PUE

$$PUE_3 = \frac{E_{\text{del,el}} \cdot w_{\text{del,tot,el}} - E_{\text{exp,el}} \cdot w_{\text{del,tot,el}} + E_{\text{ren,el}} \cdot w_{\text{del,tot,el}}}{P_{IT} \cdot w_{\text{del,tot,el}}}$$

$$= \frac{(109012 \cdot 2.461) - (48400 \cdot 2.461) + (242447 \cdot 2.461)}{(210240 \cdot 2.461)} = 1.44$$

RER-Renewable Energy Ratio

$$RER_{EP} = \frac{\begin{array}{c}E_{ren,el} + E_{ren,cool} + (w_{del,tot,el} - w_{del,nren,el}) \cdot E_{del,el} - \\ - (w_{exp,tot,el} - w_{exp,nren,el}) \cdot E_{exp,el}\end{array}}{E_{ren,el} + E_{ren,cool} + (w_{del,tot,el} \cdot E_{del,el}) - (w_{exp,tot,el} \cdot E_{exp,el})}$$

$$= \frac{\begin{array}{c}W_{PV} + Q_{C,FC} + (w_{del,tot,el} - w_{del,nren,el}) \\ \cdot (W_{GRID,del} - W_{GRID,exp})\end{array}}{W_{PV} + Q_{C,FC} + w_{del,tot,el} \cdot W_{GRID,del} - w_{exp,tot,el} \cdot W_{GRID,exp}}$$

$$= \frac{242477 + 161220 + (2.461 - 2.135) \cdot (109012 - 48400)}{242477 + 161220 + 2.461 \cdot (109012 - 48400)} = 77\%$$

Capacity Metrics

$$CC_{IT,inst} = 1 - (P_{IT}/P_{IT,inst}) = 1 - \left(\frac{24}{28.8}\right) = 17\%$$

$$CC_{IT,des} = 1 - (P_{IT,inst}/P_{IT,des}) = 1 - \left(\frac{28.8}{31.7}\right) = 9\%$$

$$CC_{fac,inst} = 1 - (P_{fac}/P_{fac,inst}) = 1 - \left(\frac{144}{172.8}\right) = 17\%$$

$$CC_{fac,des} = 1 - (P_{fac,inst}/P_{fac,des}) = 1 - \left(\frac{172.8}{172.8}\right) = 0\%$$

$$CC_{IT,m2} = 1 - \left(\frac{\text{Actual whitespace } m2 \text{ occupied}}{\text{Design withespace } m2}\right) = 1 - \left(\frac{13}{18}\right) = 28\%$$

3.6.2 Example 2. District Cooling and Heat Reuse

Example 2 corresponds to a concept where chilled water for aircooling is supplied by a district cooling system to the Data Centre. Additionally, heat from direct liquid cooling is reused for space heating by means of a heat pump (Figures 3.15 and 3.16).

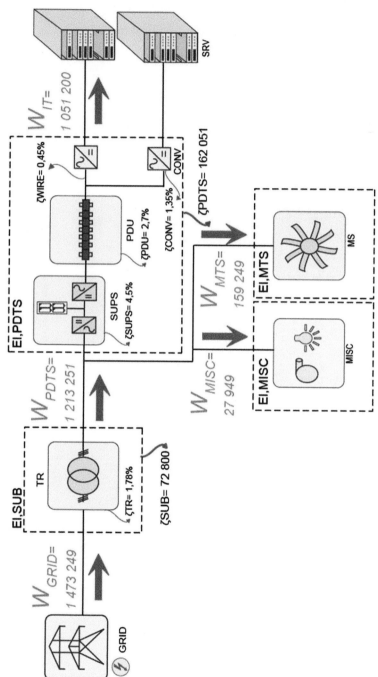

Figure 3.15 Example 2: District cooling and heat reuse, electrical scheme and main energy flow.

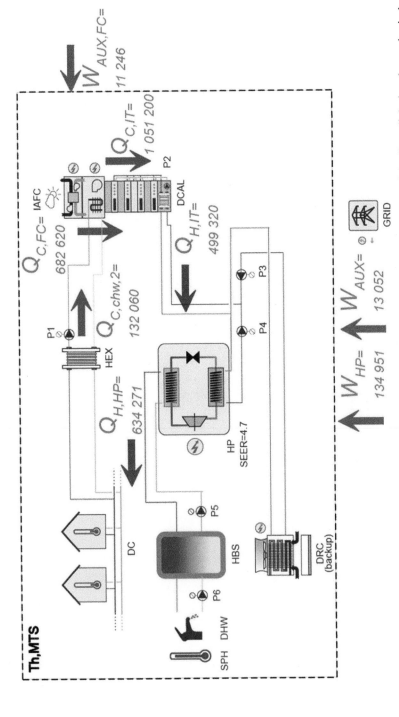

Figure 3.16 Example 2: District cooling and heat reuse, thermal scheme and main thermal and electricity flows linked to the mechanical technical systems.

Main Parameters of the System

The main figures for this example related to the IT power and main losses are similar to the one in Example 3. The location of the case is Chemnitz, and free cooling is used when it is possible. A *COP* of 4.7 is assumed for the heat pump.

Regarding the exported heat, weighting factor different from zero is considered, meaning exported heat will be subtracted in energy balances. It is assumed that the wasted heat is reused in a district heating with non-renewable energy with gas as energy source (Figure 3.17 and Table 3.4).

$$P_{IT} = 120 \text{ kW} \quad P_{fac} = 286.6 \text{ kW}$$

$$P_{IT,inst} = 1.2 \cdot P_{IT} = 144 \text{ kW}_{el}$$

$$P_{fac,inst} = 1.2 \cdot P_{fac} = 343.9 \text{ kW}_{el}$$

$$P_{IT,des} = 1.1 \cdot P_{IT,inst} = 158.4 \text{ kW}_{el}$$

$$P_{fac,des} = 1.0 \cdot P_{fac,inst} = 378.3 \text{ kW}_{el}$$

$$COP_{HP} = 4.7 \quad EER_{FC} = 60.7$$

$$W_{IT} = 1051200 \text{ kWh}$$

$$Q_{C,FC} = 682620 \text{ kWh}_{th}$$

$$Q_{C,chw,2} = 132060 \text{ kWh}_{th}$$

$$Q_{H,HP} = 634271 \text{ kWh}_{th}$$

$$Q_{C,chw,1} = Q_{H,IT} = (Q_{C,IT} - Q_{C,FC} - Q_{C,chw,2}) + \xi_{th,MTS}$$
$$= (1051200 - 682620 - 132060) + 262800 = 499320 \text{ kWh}_{th}$$

$$W_{HP} = \frac{Q_{H,HP}}{SEER} = \frac{634271}{4.7} = 134951 \text{ kWh}_{el}$$

$$W_{MTS} = W_{HP} + W_{AUX} + W_{AUX,FC} = 134951 + 13052 + 11246$$
$$= 159249 \text{ kW}_{el}$$

$$W_{AUX} = 13052 = 10183 + 2013 + 857 \text{ kWh (cooling tower +}$$
$$+ \text{ cooling distribution } + \text{ heat distribution)}$$

$$W_{GRID} = \zeta_{SUB} + W_{PDTS} + W_{MISC} + W_{MTS}$$
$$= 72800 + 1213251 + 27949 + 159249 = 1473249 \text{ kWh}$$

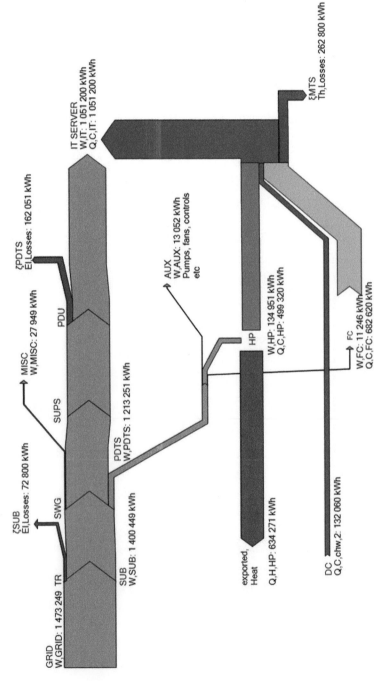

Figure 3.17   Example 1: District cooling and heat reuse, Sankey diagram.

Table 3.4 Example 3. Summary table of main energy flows

Symbol	Definition	kWh/year	PE nren	PE ren	PE tot
$E_{\text{del,el}}$	Delivered grid electricity	1 473 249	3 145 387	480 279	3 625 666
$E_{\text{ren,cool}}$	Aero thermal energy, Free cooling	682 620	0	682 620	682 620
$E_{\text{del,Dcool}}$	District cooling	132 060	26 412	105 648	132 060
$E_{\text{exp,heat}}$	Exported	634 271	634 271	0	634 271

Non-Renewable Primary Energy

$$PE_{DC,\text{nren}} = E_{\text{del,el}} \cdot w_{\text{del,nren,el}} + E_{\text{del,Dcool}} \cdot w_{\text{del,nren,DCool}} -$$

$$- E_{\text{exp,heat}} \cdot w_{\text{exp,nren,heat}} = [(1473249) \cdot 2.135 + 132060 \cdot 0.2] -$$

$$- 634271 \cdot 1.0 = 2537528 \text{ kWh}$$

Total Renewable Primary Energy

$$PE_{DC,\text{tot\_op2}} = E_{\text{del,el}} \cdot w_{\text{del,tot,el}} + E_{\text{del,Dcool}} \cdot w_{\text{del,tot,DCool}} -$$

$$- E_{\text{exp,heat}} \cdot w_{\text{exp,tot,DHeat}} = [(1473249) \cdot 2.461 + 132060 \cdot 1.0]$$

$$- 634271 \cdot 1.0 = 3806075 \text{ kWh}$$

PUE

$$PUE_3 = \frac{E_{\text{del,el}} \cdot w_{\text{del,tot,el}} + E_{\text{del,Dcool}} \cdot w_{\text{del,tot,DCool}}}{W_{IT} \cdot w_{\text{del,tot,el}}}$$

$$= \frac{1473249 \cdot 2.461 + 132060 \cdot 1.0}{1015200 \cdot 2.461} = \frac{3757726}{2587003} = 1.45$$

Renewable Energy Ratio (RER)

$$RER_{EP} = \frac{\begin{array}{c} E_{\text{ren,cool}} + (w_{\text{del,tot,el}} - w_{\text{del,nren,el}}) \cdot E_{\text{del,el}} + \\ + (w_{\text{del,tot,Dcool}} - w_{\text{del,nren,DCool}}) \cdot E_{\text{del,DCool}} \end{array}}{\begin{array}{c} E_{\text{ren,cool}} + (w_{\text{del,tot,el}} \cdot E_{\text{del,el}}) + (w_{\text{del,tot,DCool}} \cdot E_{\text{del,DCool}}) - \\ - E_{\text{exp,heat}} \cdot w_{\text{exp,tot,heat}} \end{array}}$$

$$- \frac{(w_{\text{exp,tot,heat}} - w_{\text{exp,nren,heat}}) \cdot E_{\text{exp,heat}}}{\begin{array}{c} E_{\text{ren,cool}} + (w_{\text{del,tot,el}} \cdot E_{\text{del,el}}) + (w_{\text{del,tot,DCool}} \cdot E_{\text{del,DCool}}) \\ - E_{\text{exp,heat}} \cdot w_{\text{exp,tot,heat}} \end{array}}$$

$$= \frac{\begin{array}{c} 686620 + (2.461 - 2.135) \cdot 1473249 + (1.0 - 0.2) \cdot 132060 - \\ - (1.0 - 1.0) \cdot 634271 \end{array}}{686620 + 2.461 \cdot 1473249 + 1.0 \cdot 132060 + 1.0 \cdot 634271}$$

$$= \frac{1268547}{3806075} = 33\%$$

Capacity Metrics

$$\mathrm{CC_{IT,inst}} = 1 - (P_{IT}/P_{IT,inst}) = 1 - \left(\frac{120}{144}\right) = 17\%$$

$$\mathrm{CC_{IT,des}} = 1 - (P_{IT,inst}/P_{IT,des}) = 1 - \left(\frac{144}{158.4}\right) = 9\%$$

$$\mathrm{CC_{fac,inst}} = 1 - (P_{fac}/P_{fac,inst}) = 1 - \left(\frac{286.6}{343.9}\right) = 17\%$$

$$\mathrm{CC_{fac,des}} = 1 - (P_{fac,inst}/P_{fac,des}) = 1 - \left(\frac{343.9}{378.3}\right) = 9\%$$

References

[1] (2016). Retrieved from http://projects.dc4cities.eu/projects/smart-cities-cluster

[2] 451 Research. (2012). *The economics of prefabricated modular datacenters.*

[3] AGFW. (2010). *AGFW-Arbeitsblatt FW 301 Teil 1. Energetische Bewertung von Fernwärme: Bestimmung der spezifischen Primärenergiefaktoren für Fernwärmeversorgungssysteme.* Frankfurt am Main.

[4] Allied Telesis. (2012). *Fundamental ways to reduce data center OPEX costs.* Retrieved 09 23, 2014, from http://www.slideshare.net/IPExpo/dco-day-1-0950allied-telesismelvyn-wray

[5] Brill, K. G., & Strong, L. (2013). *Cooling Capacity Factor (CCF) reveals stranded capacity and Data Centre cost savings.* upsite.com: Upsite Technologies, Inc.

[6] Cao, S., Hasan, A., & Sirén, K. (2013). On-site energy matching indices for buildings with energy conversion, storage and hybrid grid connections. *Energy and Buildings, 64,* 230–247.

[7] CEN. (2007). *EN 15459 Energy performance of buildings – Economic evaluation procedure for energy systems in buildings.*

[8] CEN/TC 371. (2016). *FprEN ISO 52000-1. Energy performance of buildings – Overarching EPB assessment – Part 1: General framework and procedures (draft).*

[9] Duffie, J., & Beckman, W. (1991). *Solar Engineering of Thermal Processes.* New York: John Wiley & Sons, Inc.

[10] Ebay. (2013). *Digital Service Efficiency – solution brief.* Retrieved 09 6, 2014, from http://tech.ebay.com/sites/default/files/publications/eBay-DSE-130523.pdf

[11] EC. (2013). *COMMISSION DECISION 2013/114/UE of 1 March 2013 establishing the guidelines for Member States on calculating renewable energy from heat pumps from different heat pump technologies.*

[12] EC. (2012a). *COMMISSION DELEGATED REGULATION (EU) No 244/2012 of 16 January 2012 supplementing Directive 2010/31/EU of the European Parliament and of the Council.* Retrieved 09 21, 2014, from http://ec.europa.eu/energy/efficiency/buildings/buildings_en.htm

[13] EC DG Connect; Unit Smart Cities & Sustainability. (2014). *Environmentally sound Data Centres: Policy measures, metrics, and methodologies. Summary report.* Brussels.

[14] EC. (2012b). *Guidelines accompanying Commission Delegated Regulation (EU) No 244/2012.* Retrieved 09 23, 2014, from http://eur-lex.europa.eu/legal-content/EN/TXT/PDF/?uri=CELEX:52012XC0419(2)&from=EN

[15] ETSI GS OEU 001. (2013). *Operational energy Efficiency for Users (OEU); Technical Global KPIs for Data Centres.* France: ETSI.

[16] Facebook. (2014). *PUE-WUE Prineville.* Retrieved 09 18, 2014, from https://www.facebook.com/PrinevilleDataCenter/app_399244020173259

[17] Future Facilities. (2014). *ACE – Five reasons your Data Center's Availability, Capacity and Efficiency are being compromised.* Retrieved 09 18, 2014, from http://www.futurefacilities.com/solutions/ace/ace_assessment.php

[18] Gavaldà, O., Depoorter, V., Oppelt, T., & van Ginderdeuren, K. (2014). *Report of different options for renewable energy supply in Data Centres in Europe.* Retrieved from http://www.renewit-project.eu/wp-content/files_mf/1399286639D4.1ReportonRESsupply.pdf

[19] Global Taskforce. (2014). *Harmonizing global metrics for data center energy efficiency, Global Taskforce Reaches Agreement Regarding Data Center Productivity.*

[20] *Green IT Promotion Council*. (n.d.). Retrieved 09 6, 2014, from http://home.jeita.or.jp/greenit-pc/e/topics/release/100316_e.html

[21] Henning, H.-M., & Wiemken, E. (2011). Appropiarte solutions using solar energy – basic comparison of solar thermal and photovoltaic approaches. *4th Internationa Conference Solar Air-conditioning.* Larnaka.

[22] Iosup, A., Ostermann, S., Yigitbasi, M., Prodan, R., Fahringer, T., & Epema, D. (2011). Performance Analysis of Cloud Computing Services for Many-Tasks Scientific Computing. *IEEE Transactions on Parallel and Distributed Systems.* doi: 10.1109/TPDS.2011.66

[23] ISO/EIC. (2016). *ISO/IEC 30134-2:2016 Information Technology – Data centres – Key Performance Indicators – Part 2: Power usage effectivness (PUE).* ISO/EIC.

[24] Jackson, K., Ramakrishnan, L., Muriki, K., Canon, S., Cholia, S., Shalf, J., et al. (2010). Performance Analysis of High Performance Computing Applications on the Amazon Web Services Cloud. *2010 IEEE Second International Conference on Cloud Computing Technology and Science (CloudCom).* doi: 10.1109/CloudCom.2010.69

[25] Klein, S., Beckman, W., & Duffie, J. (1976). A design procedure for solar heating systems. *Solar Eenergy, 18*, 113–127.

[26] Kousiouris, G., Menychtas, A., Kyriazis, D., Gogouvitis, S., & Varvarigou, T. (2014). Dynamic, behavioral-based estimation of resource provisioning based on high-level application terms in Cloud platforms. *Future Generation Computer Systems, 32*, 27–40.

[27] Monteiro, L., & Vasconcelos, A. (2013). Survey on Important Cloud Service Provider Attributes Using the SMI Framework. *Procedia Technology, 9*, 253–259.

[28] Newcombe, L. (2010). *Future of data centre efficiency metrics.* http://dcsg.bcs.org: Data Centre Specialist Group.

[29] Noris, F., Musall, E., Salom, J., Berggren, B., Jensen, S. Ø., Lindberg, K., et al. (2014). Implications of weighting factors on technology preference in net zero energy buildings. *Energy and Buildings, 82*, 250–262.

[30] Rasmussen, N. (2012). *Avoiding Costs from Oversizinig Data Center and Network Room Infrastructure. Withe Paper 37.* www.apc.com: Schneider Electric – Data Center Science Center.

[31] REHVA Report n° 4. (2013). *REHVA nZEB technical definition and system boundaries for nearly zero energy buildings.* www.rehva.eu: Jarek Kurnitski (Editor).

[32] Salom, J., Marszal, A. J., Candanedo, J., Widén, J., Lindberg, K. B., & Sartori, I. (2014, March). *Analysis of Load Match and Grid Interaction Indicators in NZEB with High-resolution Data.* Retrieved 09 25, 2014, from http://task40.iea-shc.org/data/sites/1/publications/T40A52–LMGI-in-Net-ZEBs–STA-Technical-Report.pdf

[33] Salom, J., Widén, J., Candanedo, J., & Lindberg, K. B. (2014). Analysis of grid interaction indicators in Net Zero-energy buildings with sub-hourly collected data. *Advances in Building Energy Research* (DOI: 10.1080/17512549.2014.941006).

[34] Salom, J., Widén, J., Candanedo, J., Voss, K., & Marszal, A. (2011). Understanding Net Zero Energy Buildings: Evaluation of Load Matching and Grid Interaction Indicators. *Proceedings of Building Simulation 2011, 12th Conference of IBPSA, Sydney,* 2514–2521.

[35] Sartori, I., Napolitano, A., & Voss, K. (2012). Net Zero Energy Buildings: a consistent definition framework. *Energy and Buildings* (doi:10.1016/j.enbuild.2012.01.032).

[36] Shrestha, N. L., Oppelt, T., Urbaneck, T., Càmara, Ò., Herena, T., Oró, E., et al. (2014). *Catalogue of advanced technical concepts for Net Zero Energy Data Centres. Draft version.* www.renewit-project.eu: RenewIT Project.

[37] Sisó, L., Fornós, R. B., Napolitano, A., & Salom, J. (2012, 12 28). *D5.1 White paper on Energy- and Heat-aware metrics for computing nodes.* Retrieved 09 6, 2014, from http://www.coolemall.eu/documents/10157/25512/CoolEmAll+-+D5.1+Metrics+v1.4.pdf

[38] Sisó, L., Salom, J., Oró, E., Da Costa, G., & Zilio, T. (2014, 03 31). *D5.6 Final metrics and benchmarks.* Retrieved 09 6, 2014, from http://www.coolemall.eu/documents/10157/25512/CoolEmAll+-+D5.6+Final+metrics+and+benchmarks+-+v1.0.pdf

[39] Smart City Cluster Collaboration. (2014). *Cluster Activities Task 3.*

[40] Smart City Cluster Collaboration. (n.d.). *Existing Data Centre Energy Metrics – Task 1.* Retrieved from https://renewit00.sharepoint.com/renewit/Project/WP3-Monitoring%20Systems%20and%20metrics/Smart%20City%20Cluster%20Collaboration/Deliverables/Task%201%20-%20List%20of%20DC%20Energy%20Related%20Metrics%20Final.pdf

[41] Smart City Cluster Collaboration. (2014). *Existing Data Centres energy metrics – Task 1.*

[42] Smart City Cluster Collaboration. (2014). *Smart City Cluster Collaboratio 1st WS Minutes. Barcelona.*

[43] *The Green Grid. Library & Tools.* (n.d.). Retrieved 09 6, 2014, from http://www.thegreengrid.org/library-and-tools

[44] The Green Grid. White paper #13. (2008). *A Framework for a Data Center Energy Productivity.*

[45] UpTime Institute. (2008). *A Simple Model for Determining True Total Cost of Ownership for Data Centers,.*

[46] Van der Ha, B., & Nagtegaal, B. (2014). *Data Centres: Market Archetypes and Case Studies.* www.renewit-project.eu: RenewIT Project.

[47] Whitehead, B. (2013). *Life cycle assessment of data centres and the development of a software tool for its applications.* London: PhD Thesis, London South Bank University.

[48] Wolf, M.-A. (2013). *KPIs for Green Data Centres. Background paper for stakeholder engagement and a workshop.* Berlin: maki Consulting.

[49] www.ipmvp.org. (2014). *International Performance Measurement & Verification Protocol.* Retrieved 09 21, 2014, from http://www.nrel.gov/docs/fy02osti/31505.pdf

[50] Salom, J., Garcia, A., Oró, E., Cesarini, D. N., Metrics for Net Zero Energy Data Centres. (2014), RenewIT Project, http://www.renewit-project.eu/wp-content/files_mf/1413980534D3.1MetricsforNetZeroData Centresv9.0_final.pdf

4

Advanced Technical Concepts for Efficient Electrical Distribution and IT Management

Eduard Oró[1], Mauro Canuto[2] and Albert Garcia[1]

[1]Catalonia Institute for Energy Research – IREC, Spain
[2]Barcelona Supercomputing Center – Centro Nacional
de Supercomputación, Spain

4.1 Advanced Technical Concepts for Efficient IT Management

In this section, advanced technical concepts for efficient IT management are presented which aim in reducing the electricity demand of the IT equipment due to higher performance ratios by the use of consolidating tasks. Additionally, strategies for workload shifting are addressed.

Green algorithms rely on virtualisation [1] as a core technology that simultaneously executes full operating systems as guests within a single hardware node. Virtualisation brings the following advantages for both clients and cloud providers:

1. Physical hosts can be shared transparently to the clients, which are isolated as if each client were using a dedicated physical host. Virtualisation allows common users to get administrative permissions to configure the operating system, networking, and to install and uninstall software packages.
2. The resources (i.e. CPU or Memory) can be dynamically assigned and unassigned to the virtual machines (VMs) at runtime.
3. VM can easily migrate between physical resources at runtime. The migration is transparent to the user. Migration allows distributing VMs at runtime across the resources to increase server consolidation and save energy costs.

The flexibility of cloud computing and its success as a business model involve that users with diverse workload requirements access the cloud. Tasks handled by clouds may be CPU-intensive, I/O-intensive, memory-intensive, disk-intensive or even a combination of them. The high costs of energy for current Data Centres require, in addition to an efficient building and hardware design, the addition of policies and models to assess the VM allocation and management process to minimise the energy costs also from the software perspective.

Modern approaches for energy-aware IT management rely on modelling of the power consumption of the current workloads that run within modern hardware architectures [2]. There is a noticeable difference in power consumption when the tasks dominate different resources such as CPU, memory, network and hard drive [3]. Therefore, this means that a 100% HPC load which dominates mainly CPU is not the same that a 100% data load which dominates mainly hard drive. Chen et al. [4] build a linear power model that represents the behaviour of hardware nodes running VM under high-performance computing workloads. Their model may not be suitable to perform accurate predictions since it relies in hardware nodes. Several authors [5, 6] build power models to infer power consumption that apply to VM's power metering by using existing instrumentation in server hardware and hypervisors. However, none of these approaches considers the impact of hardware heterogeneity. In addition to the hardware heterogeneity, the heterogeneity of workloads (high-performance computing, data-intensive, real-time web workloads, etc.) must be also taken into account. The energy and power models of both hardware and workloads are the basis for cloud resources allocation and management algorithms. Hypervisor-level resource management methods (VM placement, resizing and migration) may be used to improve energy saving of cloud Data Centres [7, 8]. Consolidating the maximum number of VMs within the minimum number of physical hosts while turning off the idle hosts would minimise the energy impact while maximising the energy efficiency of the Data Centres. Thus, here it can be differentiated between two IT management strategies:

- **Consolidation strategy**. The objective of this strategy is to allocate the VMs, necessaries to satisfy the IT workload demand, in the minimum number of servers. Therefore, some servers are working at full load and the rest are kept in an idle state.
- **Turn-off idle servers**. It is a complementary strategy to the consolidation method, where the servers that are not being used are turned off, instead of being in an idle state.

As an example, Figure 4.1 shows the IT load of 10 different servers for two different scenarios. The reference scenario shows random allocation of IT load in all the available servers while the consolidation scenario concentrates the overall IT load only in the minimum required servers which in this case is 5 servers at full load. In this situation, the other servers (from 6 to 10) are in idle mode. Notice that when servers are in idle mode, they still consume energy and will depend on the typology and architecture of the servers. As an example, Figure 4.2 shows the nominal energy consumption for the same servers. Due to the implementation of consolidating strategies, Data Centre can save energy but will depend on the IT workload and the idle server energy consumption if this strategy will have impact or not.

Moreover, there is the possibility to turn off the servers in idle position. Figure 4.3 shows the nominal energy consumption in this situation, and it is clear that now the potential energy reduction is higher. Notice that the numbers are only to show the concepts and, in the reality, they can change in function of the IT load, servers configuration, etc.

However, these techniques would decrease the performance and the quality of service of the deployed services [9]. A trade-off is required between energy and performance that implies distributing (as opposite to consolidating) VMs with special performance requirements. Most works about consolidation of VMs focus on performance [10] and processor energy consumption [11] but do not consider the energy consumed by VM migration [12]. In the literature, it is

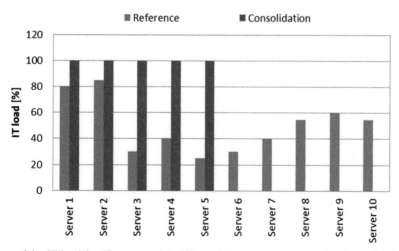

Figure 4.1 IT load for 10 servers with different IT management strategies (consolidation).

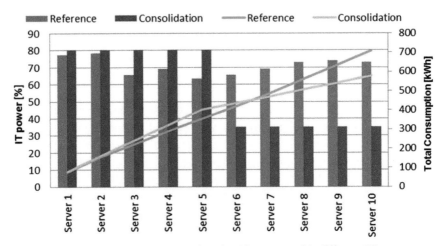

Figure 4.2 Nominal energy consumption for 10 servers with different IT management strategies (consolidation).

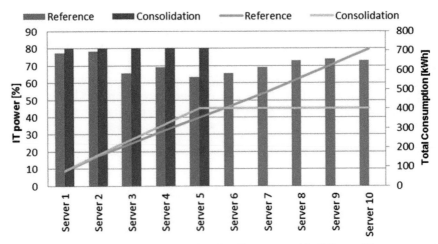

Figure 4.3 Nominal energy consumption for 10 servers with different IT management strategies (turn off servers at idle position).

available some work with potential energy savings up to 85% (but increasing the leeway by considering off-peak hours and bigger Data Centres) [13, 14]. Other approaches that are more realistic reported energy savings between 15 and 30% [15, 16]. Goiri et al. [15] measured an improvement of up to 15% when providing consolidation techniques that are aware of the power efficiency with the support for migration of VMs (Figure 4.4). The same authors showed how exploiting heterogeneity of Data Centres can improve power efficiency by up to 30% (Figure 4.5) while keeping a high level of SLA enforcement. In addition to the logical overhead caused by the overloading of tasks, the interferences between the different tasks that run in the same pool of resources must be considered [17–21]. Such interference would cause an overhead that consumes extra power and reduces energy efficiency.

When measuring the energy consumption of tasks within the resources, the thermal impact of the tasks must be considered (because the cooling systems also consume energy). It is required to calculate the relation between temperature and dissipated power for each node in order to engage thermal-aware temperatures, such as playing with "hot" and "cold" tasks to control temperature [22] or placing "hot" tasks to "cold" or best-cooled processors [23].

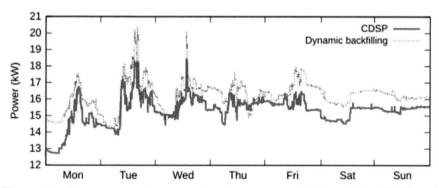

Figure 4.4 Power consumption comparison between consolidation policies (CDSP) and dynamic backfilling policy [15].

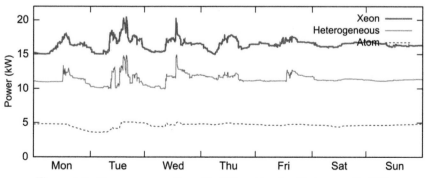

Figure 4.5 Power consumption with different levels of heterogeneity [15].

When the objective is not only to increase the energy efficiency but to maximise the ecological efficiency (reducing emissions and pollution), Data Centres that partially operate with renewable energies may schedule their workloads (if possible) according to the availability of such energies [24, 25]. For example, a Data Centre connected to solar PV schedules batch applications to the hours with the highest solar radiation. To show this, Figure 4.6 shows the IT power consumption for a Data Centre which is also connected to a PV system. Notice that during the peak PV production, the Data Centre is not consuming all the electricity generated by renewables and therefore should be sent to the main grid. On the other hand, when IT scheduling strategy is implemented in the Data Centre, the IT load can be adapted to the PV production as Figure 4.7 shows. In this situation, the Data Centre is fitted to the PV production increasing the share of renewables and decreasing the CO_2 emissions. Notice that in both scenarios, the IT consumption is the same but the execution time is different. This strategy cannot be implemented in web workload when the task should be done as a request for the user but for HPC and Data workloads exist the possibility to schedule the execution of task when it is more profitable for the Data Centre. As a complementary policy, Data Centres equipped with renewable energies that could be activated on demand (e.g. CHP or fuel cells running with biomass or biogas) could adapt energy supply to the expected workload [24].

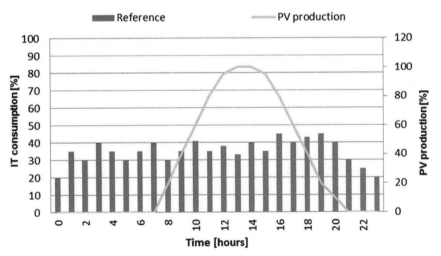

Figure 4.6 Schematic diagram of IT power consumption and PV power production without an optimisation.

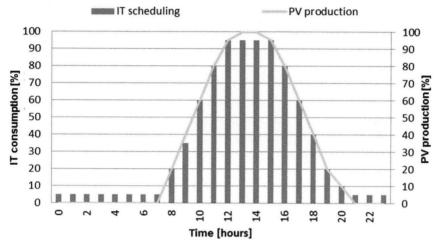

Figure 4.7 Schematic diagram of IT power consumption and PV power production under IT scheduling strategy.

Within the RenewIT project, a virtual machine manager (VMM) for Data Centre workload management has been developed to optimise the VMs placement in order to maximise the efficient usage of the resources (i.e. renewable energy) while keeping the workload performance standards that are mandatory by the cloud users. The VMM can be downloaded from the official BSC-RenewIT open source repository [26]. The main functionality of the VMM is to perform the deployment of VMs according to a policy specified by the owner of the infrastructure. Several policies have been included in the VMM:

- Distribution: It distributes the VMs trying to maximise the number of servers used. When using this algorithm, if two scenarios use the same number of servers, the one where the load of each server is more balanced is considered first. This policy is not energy saving but aims to maximise the performance and hence can be used for comparison purposes with the other policies.
- Consolidation: It distributes the VMs in a way that the number of servers that are being used are minimised.
- Energy aware: It deploys the VMs in the hosts where they consume less energy, according to the models and forecasts that were developed and released in Deliverable 2.2 of the RenewIT project [26].

In order to apply the scheduling policies described (distribution, consolidation and energy aware), the VMM needs to interact with other components such as the Energy Modeller (provides forecasts related to the energy consumption that a VM would have on a particular host) and the Infrastructure Monitor (to know the load of each host of the cluster – Ganglia [27] and Zabbix [28] – monitoring systems are used). Apart from the scheduling of VMs using the policies described above, the VMM offers other functionalities:

- It can manage the life cycle of VMs. This means that, by using the API that the VMM provides, it is possible to reboot, shutdown, suspend, restart and destroy VMs.
- It is also possible to perform queries to know at any moment the state of each of the VMs deployed and retrieve information about them: CPUs, reserved RAM and disk, IP address, the host where they are deployed, their creation date and time, etc.
- The VMM can also be used to manage the images from which VMs are instantiated. Specifically, using the VMM, it is possible to retrieve the information of all the images that have been registered, delete them and upload new ones from public URLs or from a local URI accessible by the VMM.
- Using the VMM, it is possible to retrieve information about the hosts available in the cluster where it is operating. It is possible to check the capacity of the hosts in terms of CPUs, RAM and disk as well as their current load. The VMM is also able to check the power consumption of each of the hosts at any given time.
- Finally, the VMM can be used to calculate energy estimates. Given the characteristics of a VM or a set of VMs, the VMM is able to calculate what their power consumption would be.

The main scientific contribution of the VMM developed in RenewIT project is the implementation of a scoring policy that feeds a heuristic local search algorithm that is able to improve the allocation and management of VMs to minimise the usage of energy while maximising the usage of VMs. The green algorithms developed allow setting the two already mentioned techniques: self-adaptation to enable migration of VMs at runtime and time switching to postpone batch jobs to the future. To show the VMM operative, Figure 4.8 shows the IT power consumption for two strategies, the first one is a random allocation of the IT jobs while the other is the consolidation, so the VMM distributes the VM to minimise the number of servers. Using the consolidation strategy, the total energy consumption is being reduced up to 13%. However,

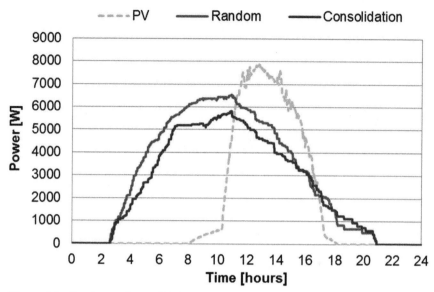

Figure 4.8 Random and consolidation strategies consumption for the same IT workload.

this strategy does not consider the availability of renewable energy sources which in this case is PV production. To greener the Data Centre, the energy-aware strategy is also implemented; here, the VMM locates the IT load to minimise the power consumption and maximise the usage of renewable energy of the Data Centre while fulfilling the performance requirements of all the VMs. Notice that there are restrictions on top of that to ensure that the tasks are always executed on time. Figure 4.9 shows the power profile of the power-aware strategy as well as the random strategy. It can be seen that the power aware is trying to follow the PV profile production over the day. In this situation, the energy consumption reduction is up to 25%, while the amount of green energy used from the PV has increased up to 60%, so the 60% of the total energy consumption under power aware is green energy.

The VMM demonstrated the feasibility of live migration for generic VMs in order to maximise the overall performance of the system in terms of energy efficiency. Moreover, to achieve the full utility of the self-adaptation policies, remote sleep/awake mechanisms must be provided to the physical nodes. This strategy will save energy when the consolidation policies allow complete freeing the load of a part of the resources.

When operating with different Data Centres that are geographically distributed, IT management policies could also minimise the energy impact and

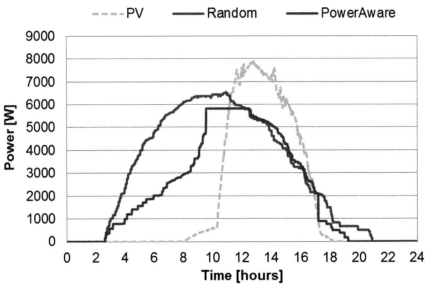

Figure 4.9 Random and power-aware strategies consumption for the same IT workload.

maximise the ecological efficiency by considering the spot status of each Data Centre [29–31]: energy cost, availability of energy according to their sources, or the status of the workloads and the physical nodes. Therefore, depending on the strategy, the Data Centre operators can manage the IT load between both infrastructures. As an example, Figure 4.10 shows the IT load in two Data Centres (A and B) as well as the percentage of renewables in the grid of each of the Data Centres. In this case, the objective is to run the IT workload using as maximum as possible the available renewable energy in the grid and therefore reducing the CO_2 emissions. Therefore, following this example is preferable to run the IT load in Data Centre A from 00:00 to 05:00 since the renewable energy ratio in the grid is higher in Grid A than in Grid B but after 05:00, it is better to move the IT load to Data Centre B. Therefore, when a geographically distributed IT strategy is implemented between Data Centre A and Data Centre B, the IT load is moved between the Data Centres, as Figure 4.11 shows. Notice that in both scenarios, the IT consumption is the same but the execution Data Centre is different.

Within the RenewIT project, also a VMM with a supra-Data Centre component has been developed that every time that a new VM is going to be deployed, it can choose the Data Centre destination in order to minimise the overall consumption or to maximise the use of on-site renewables or the

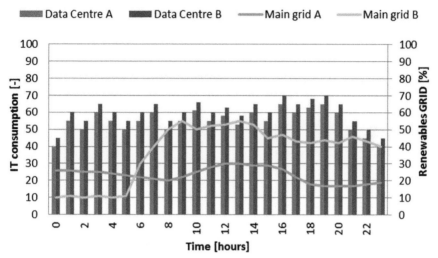

Figure 4.10 IT power consumption and percentage of renewables in the grid for 2 Data Centres.

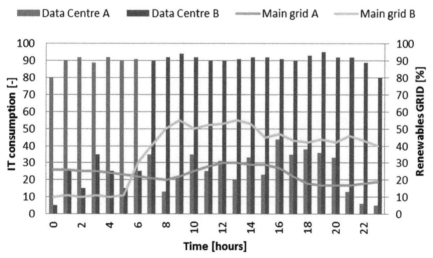

Figure 4.11 IT power consumption and percentage of renewables in the grid for 2 Data Centres under geographical IT scheduling strategy.

renewable grid ratio. As an example of the VMM deployment, Figure 4.12 shows the Data Centre power consumption of a random IT workload allocation strategy between two Data Centres. Notice that both facilities are located

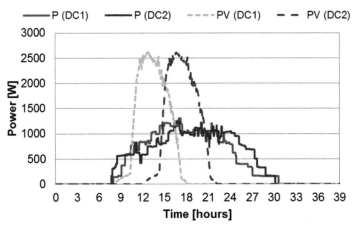

Figure 4.12   Energy consumption for an IT workload random allocation between two different Data Centres.

in different location as the PV power consumption shows (i.e. Athens and Seville). In this situation, the VMM does not considerate any relationship between the IT workload allocation and the green energy available in the facility, and therefore, the IT workload is allocated randomly in any of the two federated Data Centres. On the other hand, when the VMM is updated with the supra-Data Centre component, the VM will be deployed in one of the federated Data Centres to minimise the overall energy consumption or to maximise the use of on-site renewables. When this strategy is implemented in the same situation than explained before (Figure 4.13), the energy reduction is only being reduced by 2% but the share of renewables in the federated Data Centres is being increased up to 20%.

4.2 Advanced Technical Concepts for Efficient Electric Power Distribution

4.2.1 Introduction

The electrical distribution system of the Data Centre includes, among other equipment, switches, panels and distribution paths, UPSs and autonomous diesel generators for power backup in case of mains failure. These represent important expenditures for the Data Centre owner. However, since security of supply is principal for the installation, these expenditures are somehow unavoidable. The UPS is one of the main components of the electrical system of a Data Centre. This device is composed by an energy container (the most

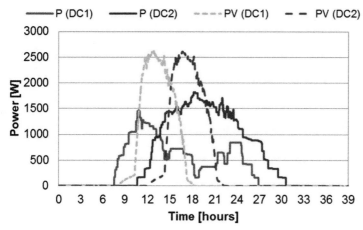

Figure 4.13 Energy consumption for an IT workload supra-Data Centre allocation between two federated Data Centres.

common is to use a battery bank or a flywheel) and a couple of power converters for its coupling to the electrical system of the facility. It is in charge of providing power to the IT equipment in case of electrical failure and before the electrical backup starts (actually a generator can operate at working conditions in less than a minute). UPS devices are one of the components of the electrical system that generates more losses. Usually, legacy UPS generates a 6% of losses compared with the total IT power consumption. Therefore, solutions able to minimise their losses have become an interesting topic in the Data Centre industry. As a result, a new generation of UPS, such as the modular or bypassed UPS, is emerging in the market. Another option for efficient electric power distribution is to enhance the energy capacity of the UPSs of the installation. This enhanced UPS could serve to provide energy during a few hours in case of a main failure as a diesel generator would provide or for peak hours. In this last scenario, the enhancement of the UPS would help to not oversize the main electrical elements to cover peak hours. However, the potential advantages of this advance electrical concept are related to the possibility of managing the energy stored in the UPS towards the technical and economic optimisation of the operation of the Data Centre. An example is to buy cheap electricity from the main grid for charging the electrical storages during the night and use it during the peak hours or when the electricity is too expensive. It is important to consider that for the economic suitability of installing energy storage systems, dedicated energy management algorithms are required. This gives the possibility to effectively time-shift the loads to low energy cost periods while still improving the power quality of the Data Centre and also being ready to act in case of a power failure in the main grid.

Therefore, in this section, a detailed behaviour of the advanced electrical concepts proposed in the framework of the RenewIT project is described. These concepts are as follows:

- Modular UPS
- Bypassed UPS in normal operating conditions
- Enhanced UPS for electrical energy storage.

4.2.2 Modular UPS

Typically, the capacity of the UPS exceeds the 100% of the maximum power of the installation and it is always connected, affecting the energy efficiency of the overall system. Here, a proposal to perform a modular design of the UPS is presented so that the number of modules connected in parallel can vary depending on the workload conditions. Each module can be activated or deactivated separately depending on workload to maximise the efficiency. The efficiency of a UPS module is particularly sensitive to workload conditions; it is minimal at low-load operation and increases until reaching its maximum at full-load operation. So adapting the number of connected modules to workload, i.e. adjusting the capacity of the UPS to the magnitude of the power to be transmitted, favours the operation of the system close to its ratings, thus maximising its efficiency. This concept is graphically depicted in Figure 4.14. Since the power is lower than 25% just one module is connected to the load while the other 3 modules are not. This obviously reduces the energy losses in those modules enhancing the overall energy efficiency.

The benefits achieved due to modularity are related with the increment of the load factor of the UPSs. The load factor is the percentage between the power demanded by the load and the nominal power of the UPS. Notice that not only the IT load profile but also relationship between the UPS design parameters and the IT installed capacity affects the instantaneous load factor. Moreover, the number of modules that the UPS will have affects the overall behaviour. After a literature research [32, 33] a UPS with 8 modules, when the UPS nominal power is between 50 and 800 kW, is used to demonstrate the feasibility of this concept.

The increment of UPS efficiency due to modularity needs to be related with the increment of the load factor between the UPS with and the one without modularity. Figure 4.15 shows the relationship between the load factor and

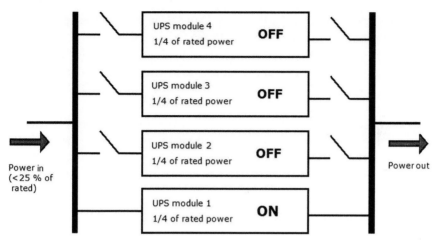

Figure 4.14 Graphical explanation for the concept of modular UPS.

the efficiency of a traditional and a modern UPS. Therefore, first it has to be calculated the overall load factor and then, using manufacturer efficiency curves, the overall energy efficiency of the UPS.

In the framework of the RenewIT project, the energy efficiency increase due to the use of modular UPSs has been done for different IT work-loads (mainly HPC and web workload) and different IT power capacities.

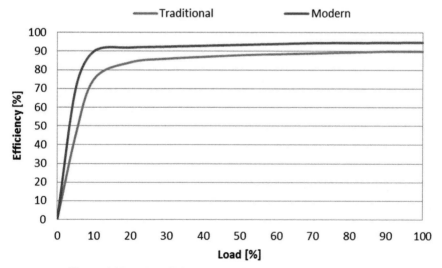

Figure 4.15 UPS efficiency depending on the load factor value.

As expected, modularity allows the UPS to work in a higher load factor and therefore increases its efficiency due to the fact that the system activates and deactivates UPS modules adjusting the nominal power close to the load demand. The results also show that the increment on the load factor in the scenarios with a web profile is lower than in the scenarios with a HPC profile. This is due to the variability of the web profile, which reaches lower load demands than in the HPC. Looking at the results in detail, the implementation of UPS modularity increases the average load factor of the UPS between 30 and 50%. This leads to an increment of 3–5% on the UPS efficiency, setting the average UPS efficiency around 97.5%. This improvement on efficiency means a reduction of 40–60% in the UPS losses, which results in 3–4% less power consumption in the overall Data Centre consumption.

4.2.3 Bypassed UPS

The majority of Data Centres utilises the static UPS in double conversion topology. In this topology, the critical loads of the Data Centre are always provided with fully conditioned mains power. Therefore, the alternating current (AC) currents and voltages from the main grid are converted to direct current (DC) and then back again to clean AC power. Applying this scheme, the losses in the UPS are noticeable because of the AC/DC double power conversion. Usually, legacy UPS generates 9% of losses compared with the total IT power consumption [34]. Therefore, solutions to minimise the UPS losses have become an interesting topic for Data Centre owners. Here, the bypassed UPS is described and analysed in detail. There are mainly three operational situations: standard situation or not bypassed, fully bypassed and partially bypassed, as Figure 4.16 shows. It can also be seen that the UPS avoids one or both converters depending on certain grid conditions and UPS characteristics, named partially and fully bypassed UPS, respectively.

Usually, UPS operates in normal conditions in order to provide the critical loads with "clean" AC power and ensure a constant power supply in case of mains failure, online operation results in no transfer time and the output voltage is of high quality. Underestimating these two main advantages, one can operate the UPS fully bypassed. In this operating mode, the power losses in the UPS are minimal since there is no power flow through the converters, but the critical loads are supplied with "dirty" power from the main grid. This reduction in power quality though can affect the operation and lifetime of the critical loads of the Data Centre and this is the main drawback of adopting this strategy. In order to improve power quality while still reducing

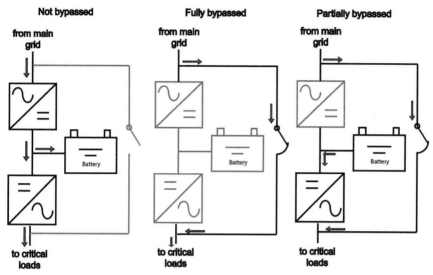

Figure 4.16 Operating modes for the UPS [35].

power losses in the UPS, it can be operated partially bypassed. In this configuration, the power flows from the main grid to the loads avoiding the AC/DC conversion in the grid side converter of the UPS; so this first power converter is offline. As a difference, the DC/AC power converter connecting the UPS to the loads is working, actively filtering the power from the main grid prior to reaching critical IT loads. As a result, this operation reduces power losses in the UPS while still improving the power quality for protection of critical loads of the Data Centre.

It is difficult to find a Data Centre operating at totally bypassed UPS due to the actual risk for this situation. Therefore, the increment of the energy efficiency of using only partially bypassed UPS in comparison with standard operational situation has been studied. The study is performed over different scenarios where different profiles and Data Centre redundancy levels are assumed. These profiles are web and HPC profile. The redundancy levels are as follows: redundancy level I, which means no redundancy in the electrical system, and redundancy level III, which means N+1 redundant component and two lines of electrical distribution, according to [36]. Moreover, in each scenario, IT power varies from 50 to 3,000 kW. On one hand, depending on the UPS topology, the transfer time between normal conditions and partially bypassed mode will be higher or lower. It is crucial to know the transfer time in order to ensure that the UPS will be able to provide power in case of main grid failures or clean the power in case of voltage disturbances, avoiding power interruptions in the IT equipment. A key factor during this process is the design ride-through of the IT power supply unit (PSU). PSU, in IT systems, can store small amounts of energy in capacitors and thereby allow a certain amount of ride-through time, which is the length of time that the IT equipment can continue to function during a complete loss of power. Usually, the ride-through time of a PSU ranges between 10 and 50 ms. Thus, it is imperative to know the ride-through capabilities of every PSU that will be powered by the UPS and know the transfer time of the UPS that needs to be lower than the PSU ride-through time. The green grid provides a summary table, for different UPS topologies, about the transfer time. This classification can be also found in the standard IEC 62040-3 [37]. On the other hand, to work with partially bypassed UPS, the grid quality also has to be considered. Even if nowadays some devices measure the quality of the power supplied by the grid, there is not too much information available in literature about the areas and the annual number of hours with an adequate grid power quality. Based on the specifications of some UPS manufacturer such as ENERGY STAR [38] and the information about grid power quality in the U.S., developed by the EPRI [39], a set of profiles showing the percentage of the annual hours working in partially bypassed mode are developed. Therefore, 3 scenarios have been considered, with the UPS working in partially bypassed mode at 25%, 75% and 95% of the annual hours.

The results of the study show that depending on the scenario (mainly IT workload and IT capacity installed) and the number of hours applying partially bypassed mode, the UPS efficiency can be improved between 0.5 and 3.0%. The highest UPS efficiency found was 98.2%, and the average UPS efficiency at partially bypass mode is around 97.5%.

4.2.4 Enhanced UPS for Electrical Energy Storage

The objective of this strategy is to enhance the capacity of the electrical energy storage in the Data Centre to optimise the energy flows into the facility. Most of the Data Centres have almost constant power consumption but in some cases, big differences can occur due to user's behaviour. In this situation, the enhancement of the electrical energy storage can really help to peak load shifting and therefore not only reduce the total capacity of the electrical elements and therefore reduce the investment cost but also to increase the energy efficiency of the system by working at higher load. Here, the enhancement of the electrical energy storage is aimed for electricity trade-off. That means if the electricity price is low, it is stored to be used when the electricity price is high. However, each scenario should be analysed in detail since the boundary conditions, in particular the electricity price, should change drastically between one and other Data Centre. In order to show the feasibility of this concept different IT capacities are studied (from 50 to 3,000 kW) and the following assumptions have been made:

- The UPS which are lead-acid batteries will only use the enhanced capacity to do smart trading. This enhancement can be up to 2.5 times the original capacity. They have a maximum nominal power of 709 kW, in function of the manufacturer specifications.
- Two different IT workload profiles have been analysed: web and HPC profile as Figure 4.17 shows.
- It is considered two scenarios for redundancy, a Tier I and a Tier III. On one hand, defining and enhancement of 100% in a 50 kW IT Data Centre with no redundancy (Tier I), the energy capacity of the UPS will be enhanced by 50 kW. On the other hand, for a 50 kW IT Data Centre with redundancy (Tier III), the total energy capacity of the UPS will be also enhanced by 50 kW, but there will be 2 UPS (line A and line B of the Data Centre, as shown in Figure 4.18.
- The electricity cost profiles are from Spanish industrial consumers (Figure 4.19).
- The economical evaluation is done in a 3-year scenario due to the fact that second life batteries should be replaced every three years of operation.
- It is assumed that the extra batteries are second life batteries, with an approximated cost of 150 €/kWh.

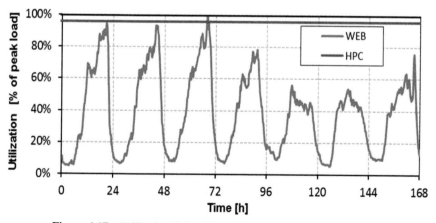

Figure 4.17 Utilisation defined by the IT load profiles web and HPC.

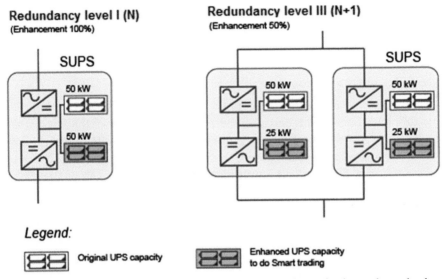

Figure 4.18   Enhancement of a UPS capacity to do smart trading, redundant and no redundant scenario [35].

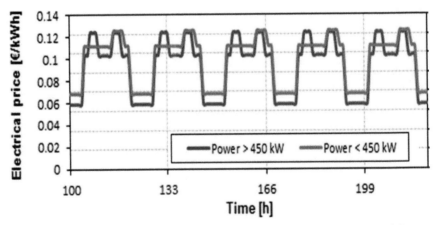

Figure 4.19   Electrical price profile for industrial consumers in Spain, five days of the year.

The smart trading algorithm is mainly governed by two conditions and three equations. The first condition compares the actual and the average electricity price. While the actual electricity price depends on the contract between the Data Centre and the electrical operator, the average is calculated using the electrical price of the last 24 hours. Depending on the comparative between the average and actual electrical price, the smart trading algorithm sends the order of charging or discharging to the UPS. If the actual price is higher than the average price, the smart trading algorithm will enforce the UPS to discharge the batteries. Otherwise, if the actual price is lower than the average, the smart trading algorithm will enforce the UPS to charge the batteries. The second condition is that the Data Centre energy management system cannot inject electricity (stored previously) to the grid. Therefore, during a discharge period, the state of charge imposed by the smart trading algorithm to the UPS will be restricted by the amount of energy that the UPS can deliver and the amount of power that the load is requiring. As example, if more power is demanded than the UPS can deliver, the state of charge becomes the minimal state of charge that the UPS can achieve, which is equivalent to discharge the extra batteries added at the UPS. Otherwise, if the UPS can deliver more power than demanded to cover the load, the smart trading algorithm calculates the exact state of charge necessary to satisfy the demand. Figure 4.20 represent the logical sequence followed by smart trading algorithm.

Where Equation (4.1) (minimal SOC), Equation (4.2) (Pmax,dch) and Equation (4.3) (reference SOC) are as follows:

$$\text{Minimal SOC} = \frac{100}{(\text{Enhancement} + 1)} \tag{4.1}$$

$$P\text{max}_{\text{dch}} = \frac{\left(\frac{100}{(\text{Enhancement}+1)} - \text{SOC}_{\text{bat}}\right) \cdot Kp \cdot V}{1000} \tag{4.2}$$

$$\text{MSOC}_{\text{ref}} = \frac{\text{PIT} \cdot 1000}{Kp \cdot V} + \text{SOC}_{\text{bat}} \tag{4.3}$$

The results of the study showed that with the boundary conditions used, especially the electricity price and the cost of the battery (150 €/kWh), there is no economic benefit of implementing extra energy storage in the system. These results do not mean that the concept is not working, it is just a matter of where to place the Data Centre and therefore which electricity cost is used, especially regarding the difference between day and night prices. In order to demonstrate this, it is calculated the maximum price of the second life battery in which the advanced energy efficiency strategy is starting to generate economic profits. For the current situation (Spain), this price is 73 €/kWh. Notice also that following the initial assumptions, not all the energy stored during a cheap electricity period could be later discharged when the prices are higher, in case that the battery could not be fully discharged before the control signal changes

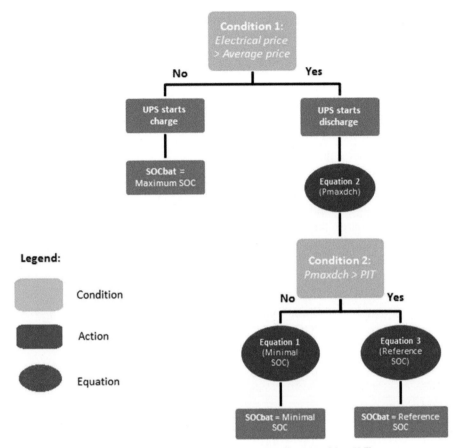

Figure 4.20   Block diagram of smart trading algorithm [35].

the UPS operational mode to charge. Additionally, when the UPS is storing energy, the IT load demand must also be satisfied. Thus, the amount of energy that can be stored by the batteries is limited because of the converter upstream of the battery. Therefore, the enhancement of the converter should also be taken into consideration for optimised strategies.

References

[1] Barham, P., Dragovic, B., Fraser, K., Hand, S., Harris, T., Ho, A., Neugebauer, R., Pratt, I., and Warfield, A. (2003). Xen and the art of virtualization. *ACM SIGOPS Operat. Syst. Rev.*, 37 (5), 164–177.

[2] Rivoire, S., Ranganathan, P., and Kozyakis, C. (2008). A comparison of high-level full-system power models, conference on Power–Aware computing and Systems (Hot Power '08)," p. 3.

[3] Smith, J. W., and Sommerville, I. (2011). "Workload classification and Software Energy Measurement for Rfficient Scheduling on Private Cloud Platforms," in *1st ACM Symposium on Cloud Computing (SOCC)*, October 2011.

[4] Chen, Q., Grosso, P., Van Der Veldt, K., De Laat, C., Hofman, R., and Bal, H. (2011). "Profiling energy consumption of VMs for green cloud computing," *IEEE 9th International Conference on Dependable, Autonomic and Sedure Computing (DASC)*, pp. 768–775, December 2011.

[5] Kansal, A., Zhao, F., Liu, J., Kothari, N., and Bhattacharya, A. (2011). "Virtual machine power metering and provisioning," *1st ACM Symposium on Cloud Computing*, pp. 39–50, October.

[6] Bohra, A., Chaudhary, V. and Meter, V. (2010). "Power modelling for virtualized clouds, *IEEE International Symposium on Parallel and Distribution Processing, Workshops and PhD Forum (IPDPSW)*, pp. 1–8, April 2010.

[7] Bo, L., Jianxin, L., Jinpeng, H., Tianyu, W., Qin, L., and Liang, Z. (2009). "EnaCloud: an energy-saving application live placement approach for cloud computing environments," *International Conference on Cloud Computing*.

[8] March, J. L., Sahuquillo, J., Petit, S., Hassan, H., and Duato, J. (2011). Real-time task migration with dynamic partitioning to reduce power consumption. *Actas de las XXII Jornadas de Paralelismo*, 185–190, September 2011.

[9] Ye, K., Huang, D., Jiang, X., Chen, H., and Wu, S. (2010). Virtual machine based energy-efficient data center architecture for cloud computing: A performance perspective. *International Conference on Green Computing and Communications and International Conference on Cyber, Physical and Social Computing*, 2010.

[10] Lu, L., Zhang, H., Smirni, E., Jiang, G., and Yoshihira, K. (2013). Predictive VM consolidation on multiple resources: beyond load balancing, *21st International Symposium on Quality of Service*, 2013.

[11] Deng, W., Liu, F., Jin, H., Liao, X., Liu, H., and Chen, L. (2012). Lifetime or energy: consolidating servers with reability control in virtualized cloud dataceneters, *4th International Conference on Cloud Computing Technology and Science*, 2012.

[12] Beloglazov, A., and Buyya, R. (2012). Optimal online deterministric algorithms and adaptive heuristics in Cloud data centers. *Concurrency Comput. Practice Exp.*, 24 (13), 1397–1420, September 2012.

[13] Uddin, M., and Rahman, A. A. (2010). Server consolidation: An approach to make data centers energy efficient & green. *Int. J. Sci. Eng. Res.*, 1 (1), October 2010.

[14] VMWare, "How VMware Virtualization Right-sizes IT Infrastructure to Reduce Power Consumption," 2011.

[15] Goiri, Í., Berral, J. L., Fitó, J. O., Julià, F., Nou, R., Guitart, J., Gavaldà, R., and Torres, J. (2012). Energy-efficient and multifaceted resource management for profit-driven virtualized data centers. *Future Generation Comput. Syst.*, 28 (5), 718–731.

[16] Beloglazov, A., and Buyya, R. (2014). OpenStack neat: A framework for dynamic and energy-efficient consolidation of virtual machines in OpenStack clouds. *Concurrency and Comput. Practice Exp.* [online].

[17] Mars, J., and Tang, L. (2013). Whare-map: heterogeneity in "homogeneous" ware house scale computers, *40th ACM/IEEE International Symposium on Computer Architecture (ISCA)*, pp. 619–630, June 2013.

[18] Mars, J., Tang, L., Hundt, R., Skadron, K., and Sofia, M. L. (2011). "Bubble-up – increasing utilization in modern warehouse scale computers via sensible co-locations," *44th IEEE/ACM Symposium on Microarchitecture (MICRO'44)*, pp. 248–259, 2011.

[19] Koh, Y., Knauerhase, R., Brett, P., Bowman, M., and Wen, Z. P. (2007). "Ana analysis of performance interference effects in virtual environments," *IEEE International Symposium on Performance Analysis of Systems and Software (ISPASS)*, pp. 200–209, April 2007.

[20] Somani, G., and Chaudhary, S. (2009). "Application performance isolation in virtualization," *IEEE International Conference on Cloud Computing (CLOUD'09)*, pp. 41–48, September 2009.

[21] Kousiouris, G., Cucinotta, T., and Varvarigou, T. (2011). The effects of scheduling, workload type and consolidation scenarios on virtual machine performance and their prediction through optimized artificial neural networks. *J. Syst. Software*, 84 (8), 1270–1290, August 2011.

[22] Yang, J., Zhou, X., Chrobak, M., Zhang, Y., and Jin, L. (2008). "Dynamic thermal management through task scheduling," *IEEE International Symposium on Performance Analysis of Systems and Software (ISPASS'08)*, pp. 190–201, April 2008.

[23] Vanderster, D. C., Baniasadi, A., and Dimopoulos, N. J. (2007). Exploiting task temperature profiling in temperature-aware task scheduling for computational clusters. *12th Asia – Pacific Conference on Advances in Computer Systems Architecture (ACSAC'07)*, pp. 175–185.

[24] Li, C., Zhou, R., and Li, T. (2013). Enabling distributed generation powered sustainable high performance data centre. *International Symposium on High-Performance Computer Archhitecture (HPCA)*, pp. 35–46, February.

[25] Goiri, I., Le, K., Haque, M. E., Beauchea, R., Nguyen, T. D., Guitart, J., Torres, J., and Bianchini, R. (2011). Greenslot: scheduling energy consumption in green datacenters. *International Conference for High Performance Computing, Networking, Storage and Analysis (SC'11), Article No. 20*, 2011.

[26] Macías, M., Canuto, M., Ortiz, D. and Guitart, J. (2015). Green management of data centres: model for energy and ecological efficiency assessment. RenewIT deliverable number 2.2.

[27] "Ganglia Monitoring System," [Online]. Available: ganglia.sourceforge. net

[28] "Zabbix: An enterprise-class open source distributed monitoring solution," [Online]. Available: www.zabbix.com

[29] Garg, S. K., Yeo, C. S., Anandasivam, A., and Buyya, R. (2011). Environment-conscious scheduling of HPC applications on distributed Cloud-oriented data centers. *J. Parallel Distributed Comput.* 71 (6), 732–749, June 2011.

[30] Pierson, J. M. (2011). Green task allocation: Taking into account the ecological impact of task allocation in clusters and clouds. *J. Green Eng.* 129–144, January 2011.

[31] Elmroth, E., Tordsson, J., Herandez, F., Ali-Eldin, A., Svard, P., Sedaghat, M., and Li, W. (2011). Self-management challenges for multi-cloud architectures. *4th European Conference on Towards a Service-Based Internet*, pp. 28–49, 2011.

[32] 2015. [Online]. Available: http://www.stat.osu.edu/~comp_exp/jour. club/CamCarSal_EngModellingSoftware-2007.pdf

[33] Chloride Trienergy, Emerson, Modular architecture for high UPS Systems, White Paper, 2011.

[34] Rasmussen, N. (2007). Electrical efficiency modeling for data centres. White Paper #113, APC, Schneider, 2007.

[35] N. E. A. Shrestha, "Deliverable D4.5 Catalogue of advanced technical concepts for Net Zero Energy Data Centres. Final version," 2015.

[36] Uptime institute, "Data center site infrastructure Tier standard: Topology," 2010.

[37] International Standard, IEC 62040-3, "Uninterruptible power systems (UPS), PART 3: Method of specifying the performance and test requirements, First edition, 1999–2003," no. First edition, 1999–2003.

[38] ENERGY STAR, *Program Requirements for Uninterruptible Power Supplies (UPSs), Test Method final*, May 2012.

[39] EPRI, "Distribution System Power Quality Assessment: Phase II – Voltage Sag and Interruption Analysis," Palo Alto CA, 2003.

[40] "OptaPlanner documentation," [Online]. Available: http://www.opta planner.org/learn/documentation.html

5

Advanced Technical Concepts for Low-Exergy Climate and Cooling Distribution

**Nirendra Lal Shrestha, Thomas Oppelt,
Verena Rudolf and Thorsten Urbaneck**

Chemnitz University of Technology, Professorship Technical
Thermodynamics, Germany

5.1 Introduction

The objective of this chapter is to describe energy efficiency strategies and advanced technical concepts for Low-Ex climate control with the aim of increasing the return temperature and the temperature difference between the air/water inlet and outlet. Proposed concepts are currently applicable best practices and novel concepts, which can improve supply and distribution of cooling in Data Centres. A quantification of the energetic savings and estimated payback period is presented for each of the concepts which are as follows:

- **Free cooling (Section 5.2):** In these strategies, the cooling demand of the Data Centre is reduced due to the utilisation of the available natural resources as a source of cooling energy.
- **Increasing the maximum allowable temperatures for IT equipment (Section 5.3):** Increasing the Delta T (air temperature difference between inlet and outlet) through the IT equipment directly reduces the required airflow rate through the whitespace. Therefore, energy required by the fans is reduced for circulating the airflow.
- **Hot or cold aisle containment (Section 5.4):** It consists to separate the IT room in hot and cold corridors for a better air management. The objective of these strategies is to avoid that air streams were affected by different phenomena such as bypass, recirculation and pressure air drop,

decreasing cooling efficiency and creating vicious cycle of rise in local temperature.

- **Variable airflow (Section 5.5):** In this concept, the airflow supply into the whitespace is adjusted depending on the required cooling load.
- **Partial loads, component oversizing and using redundant components (Section 5.6):** Using components in partial load can lead to an increasing energy efficiency. It might also be beneficial to use redundant and/or oversizing components.
- **Use of high energy efficiency components (Section 5.7):** Using highly energy efficient components can also lead to the significant reduction in the total energy demand.

5.2 Free Cooling

Free cooling is a cooling design principle, which covers a wide-spread implementation of cooling from natural resources. Free-cooling technologies can be roughly divided into the following:

- Airside free cooling: Makes use of outside air for cooling Data Centres

  - Direct airside free cooling: Drawing the cold outside air directly into the Data Centre (mixing with return air to fulfil the inlet requirements).
  - Indirect airside free cooling: Operating through air-to-air heat exchangers.

- Waterside free cooling: Utilises natural cold source through a chilled water infrastructure (air, ground, river, sea, etc.)

  - Direct water-cooled system: Natural cold water is used to cool the Data Centre. A coolant transfers heat to the seawater (river, ground) through a heat exchanger.
  - Air-cooled system: An air cooler (dry cooler) is used to cool the chilled water circulating through CRAHs when the wet-bulb temperature of the outside air is low enough.
  - Cooling tower system: A cooling tower is used to cool the chilled water circulating through CRAHs and heat exchangers. Two water loops are needed: a cooling (external) water loop and a chilled (internal) water loop.

The possibilities of free cooling depend on climatological and geographical aspects. Therefore, the location of the Data Centre defines the possibilities of applying free cooling.

Table 5.1 shows the annual hours which are suitable for direct air free cooling at different locations around Europe. As can be seen from the Table, the number of hours depends on the Data Centre operational parameters (inlet air temperature to the IT room, also see next section).

As an example, Figure 5.1 shows the estimated reduction in energy required for cooling which can be reached by applying free cooling at Barcelona. Please be aware that the tool [1] used for obtaining the data assumes hot aisle containment for the case of airside free cooling, but no aisle containment for the case of waterside free cooling. Thus, comparability of these two different concepts is limited.

Additionally, a constant chilled water set point of 7.2°C is assumed for waterside free cooling which limits the number of free cooling hours significantly. In practice, higher values are possible especially if the air supply temperature is allowed to exceed 18°C.

Table 5.1 Hours of direct air free cooling for different locations depending on the requested air supply temperature

	Data Centre air Supply Temperature				
Location	18°C	20°C	22°C	24°C	26°C
Barcelona (Spain)	4,776	4,918	5,014	5,016	5,072
Chemnitz (Germany)	6,401	6,533	6,612	6,639	6,649
Luleå (Sweden)	7,537	7,651	7,677	7,685	7,685

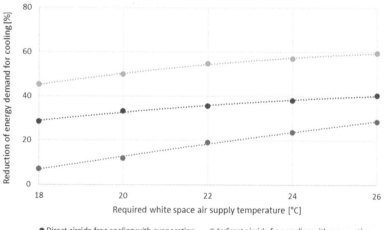

● Direct airside free cooling with evaporation ◉ Indirect airside free cooling with evaporation
● Waterside free cooling (cooling tower)

Figure 5.1 Reduction in energy demand for cooling due to free cooling in Barcelona [1].

However, it is obvious from Figure 5.1 that free cooling saves a considerable amount of energy compared to cooling by means of chillers even at the relatively warm location of Barcelona.

Due to applying free cooling, the *PUE* may generally decrease from approximately 1.8 to approximately 1.2. This results in 75% electrical savings for the mechanical installations, which implies a saving of 33% of the Data Centre's total energy demand. The payback period depends on the type of free-cooling technology, but should normally be 1–3 years.

5.2.1 Free Cooling with Direct Ambient Air

General Description

A direct air-cooling system uses outside air for cooling the Data Centre by supplying the external air directly into the Data Centre. The outside air has to be filtered before being supplied to the Data Centre. It is also possible to implement humidification and dehumidification in addition to the direct air cooling system. As a result, it is possible to use only the fan-filter system during a limited period of the year (without additional airconditioning).

During the cold season of the year, the outside air has to be humidified and heated. The reason is to minimise the risk of any damage to the server racks, caused by static electricity. Heating of the outside air is achieved by mixing a part of the recirculation air from the Data Centre with the cold outside air (instead of discharging totally the warm air to the outside). However, the supplied (mixed) air has to be humidified.

During extremely cold outside temperatures, it is necessary to recirculate the air completely. Cooling will be done by means of cooling coils (chilled water). The advantage of using chilled water cooling is a chilled water cooling plant operating 100% on free cooling (no compression cooling with the use of a refrigerant). In addition, the humidification will be considerably less. The disadvantage is the need of an additional cooling system (air-cooled chillers with free cooling mode).

Free cooling units are located adjacent to the white space and to the outside air (facade) and consist of the following components (in the sequence of the direction of the air):

- Air inlet section, including grilles and damper
- Filter section
- Mixing section (for return air)
- Fan section, including sound attenuators

- Return air section, including dampers
- Air outlet damper

In addition, humidification and dehumidification sections can be incorporated in the free cooling units.

Airside free cooling uses outside air for cooling the infrastructure. For this type of cooling, no (chilled) water is required: only air-cooling is applied (direct air or indirect air-to-air). The cooling efficiency of airside free cooling is especially high at locations with low annual average temperatures and low peak temperatures in summer.

Direct Airside Cooling

The available hours of direct airside free cooling depend on the following:

- Temperature of the outside air
- Level of humidity of the outside air
- Data Centre operating requirements

In principle, when the outside temperature is below the required temperature and the air humidity is within the ASHRAE guidelines, direct airside free cooling can be applied. This means that the amount of annual hours with direct airside free cooling can be divided into the following situations (see also Figure 5.2):

- 1A: If the temperature of the outside air is below the specified Data Centre supply temperature and the absolute humidity is within the acceptable range, supply air is composed from mixing recirculated air and outside air.

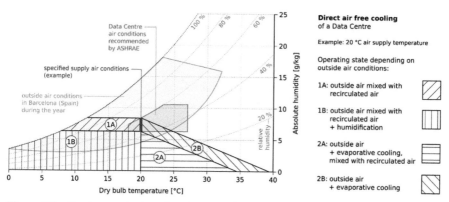

Figure 5.2 Psychrometric chart showing operating states of direct air free cooling depending on outside air conditions.

- 1B: As 1A, but with humidification of the supply air because the humidity of the outside air is too low.
- 2A: Outside air is cooled by evaporative cooling because its dry-bulb temperature exceeds the specified Data Centre supply temperature. Additionally, the outside air is mixed with recirculated air.
- 2B: As 2A, but no mixing with recirculated air is necessary because the temperature and the humidity of the cooled outside air meet the requirements of the white space.

The need for and the amount of evaporative and chiller units depend on the geographic location. At warm regions, both types of cooling are necessary while at colder regions, low temperature outside air and evaporative cooling might be sufficient. However, the relative humidity of outside air determines the applicability of evaporative cooling.

An important point is the heat generation by the supply fan. This will rise the temperature of the outside air. For example, with a temperature difference of 10 K between supply and return temperature of the Data Centre, the fan power will be about 3–5% of the IT power. This will result in a temperature rise of approximately 0.3–0.5 K. With a required white space temperature of 20°C, the maximum required outside temperature then should be 19.5–19.7°C.

Cooling Scheme

The cooling of a Data Centre by means of direct aircooling is shown in Figure 5.3.

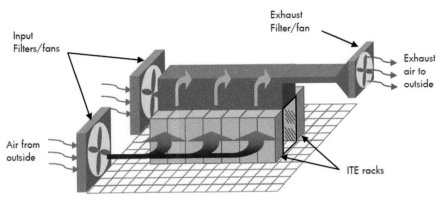

Figure 5.3 Principle of cooling a Data Centre with direct aircooling (without mixing duct) (DEERNS).

Control

During winter, the outside air is mixed with the return air from the white space. When the humidity of the mixed air is below the required humidity, extra evaporative humidification has to be applied. In this case, both the amount of outside air and the amount of humidification have to result in the required supply air temperature and humidity to the white space. Furthermore, in free cooling operation (moderate temperatures), when the evaporative cooling results in a lower supply temperature than required, first the outside air has to be mixed with the return air from the white space, before evaporative cooling is applied. At high outside temperatures, no outside air is used. The return air is cooled by means of a chiller (DX or chilled water).

Limits of Application

There are critical limits of this application, especially concerning humidity control and air quality. A precondition for successfully using direct aircooling is to allow large differences of temperature and humidity in the Data Centre. Otherwise, this concept requires a large chiller system. Since these chillers have minimal operating hours, a decentralised and modular system is possible, keeping system simplicity. However, this increases maintenance costs and results in a higher "day one" investment.

Due to direct provision of very large outside air volumes to IT equipment, the system is very vulnerable for effects from outside (e.g. external smoke, air pollution). Therefore, there is a significant risk if outside air contains harmful contaminants or particles.

Economic Aspects

There are different types of direct air units available in the market, varying from complete air handling units (or containerised units) to fan systems integrated in the building. Generally, all systems consist of filter-fan units, recirculation air and additional cooling and (de)humidification.

Costs are comparable with air handling units. Based on quotations from different suppliers of air handling units, the estimated cost is as follows (sized by m³/h airflow):

$$C = 20 \cdot \dot{V}^{0.75} \quad \text{with} \quad \dot{V} \in [5,000 \ldots 100,000 \text{ m}^3/\text{h}] \tag{5.1}$$

where C and $\dot{V}$ represent the investment cost in € and the airflow in m³/h, respectively.

If the starting point of the temperature difference between inlet and outlet temperature of the Data Centre is 12 K (= 4 Watt cooling per 1 m³/h airflow), this results in the following cost per kW cooling:

$$c_{\dot{Q}} = 1257.4 \cdot \dot{Q}_0^{-0.25} \quad \text{with} \quad \dot{Q}_0 \in [20 \dots 400 \text{ kW}] \qquad (5.2)$$

where $c_{\dot{Q}}$ and $\dot{Q}_0$ represent the specific investment cost in € /kW and the cooling capacity in kW, respectively.

5.2.2 Free Cooling with Indirect Ambient Air

General Description

An indirect air-cooling system uses outside air for cooling the Data Centre. The difference with respect to a direct air cooling system is that the external air is not supplied directly into the Data Centre. This is achieved by means of a separation between the external air and the cooling air inside the Data Centre. The hours of indirect airside free cooling can be divided into two situations:

- Air-to-air cooling
- Additional evaporative cooling (primary side) with air-to-air cooling

There are mainly two principles of indirect air-cooling:

- Using heat recovery wheels
- Using plate heat exchangers (cross flow)

Air-to-air free cooling is also directly related to the outside temperature. However, the use of a heat exchanger between the two airflows results in a higher supply temperature to the Data Centre in relation to the outside air temperature. The temperature difference between the two sides of the heat exchanger is approximately 2–3 K (depending on the design of the heat exchanger).

The fans of the outside and Data Centre airflow can be located in the warm/return airstream. This way, the heat generation of the fans does not affect the supply temperature to the white space.

Since the outside and Data Centre airflows are completely separated, free cooling can be applied for all (suitable) outside temperatures, regardless of the level of humidity of the outside air. As a result, the maximum outside (dry bulb) temperature for free cooling to achieve the required supply temperature without evaporative cooling is 2–3 K below the supply temperature to the

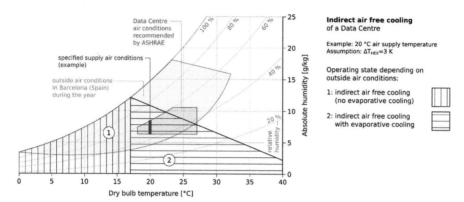

Figure 5.4 Psychrometric chart showing operating states of indirect air free cooling depending on outside air conditions.

Data Centre. An example for the suitable outside air conditions in Barcelona is shown in area 1 of Figure 5.4.

In addition to indirect air-to-air cooling, evaporative cooling can be incorporated. This makes it possible to use outside aircooling with higher outside temperatures without using a chiller. As a result, the (additional) usage of evaporative cooling increases the annual number of free-cooling hours.

When using evaporative cooling, it is possible to cool the outside air to the wet-bulb temperature. So with evaporative cooling, the maximum allowable outside air temperature is based on the wet-bulb temperature. Due to the use of a heat exchanger, the maximum outside wet-bulb temperature for free cooling to achieve the required supply temperaturethen will be 2–3 K below the supply temperature to the white space.

The area 2 in Figure 5.4 shows the outside air conditions in Barcelona which are suitable for indirect airside free cooling with evaporative cooling for the example of 20°C Data Centre supply temperature.

There are different concepts and suppliers available, for example the Green Cooling for Data Centres, or GCDC © concept. The indirect air system is equipped with a number x of fans for the internal airflow and a number y of fans for the external (outside) airflow of the unit, both as a $N + 1$ redundant configuration. The outside air fans operate on full load or partial load, depending on the outside air temperature. This way, the required electrical power of the fans is reduced to an optimum at any lower outside air temperatures. It is possible to make use of redundant components inside the indirect cooling unit. It is also an option to use extra (redundant) units to achieve certain redundancy level.

The GCDC ©️ concept is based on using heat exchangers, cooling the warm return air from the Data Centre by means of outside air. All waste heat is transferred to the outside air.

As already explained before, if sensible temperature of the outside air becomes too high to cool the Data Centre to the required or acceptable temperature (during warm days in summer), additional cooling can be applied to the outside air. This need for additional cooling therefore depends on the location of the Data Centre and the associated climate conditions. This results in the following situations:

- Cold regions without extreme warm conditions:
    - No additional cooling required
- Temperate regions with warm summer conditions (however no extreme conditions):
    - Additional cooling by means of evaporative cooling
- Warm regions with extreme summer conditions:
    - Additional cooling by means of evaporative cooling
    - Mechanical cooling (chilled water or direct expansion)

The GCDC ©️ units (Figure 5.5) consist of two separate airflows:

- *Recirculation air of the Data Centre*: The Data Centre is cooled by cold air supplied from the GCDC ©️ unit. It absorbs the heat of the ICT equipment and flows back to the GCDC ©️ unit.
- *Outside airflow*: The outside air is distributed to the heat exchanger of the GCDC ©️ unit and delivers the required cooling to the recirculation air of the Data Centre.

The GCDC ©️ units are located adjacent to the white space. The system can be incorporated on a double-storey building and is compatible for multiple-storey building configurations. The units consist of two sections with the following components (in the sequence of the direction of the air):

- Outside air section:
    - Air inlet section, including damper
    - Filter section
    - Mixing section (bypass)
    - Adiabatic cooler

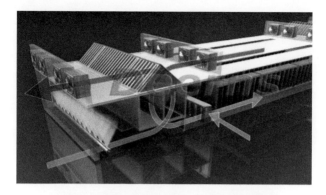

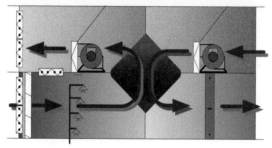

Figure 5.5 GCDC © concept – indirect air cooling using heat exchangers (DEERNS).

- Plate heat exchanger (cross flow)
- Heat exchanger chiller
- Fan section (three fans), including sound attenuators
- Bypass section, including damper
- Air outlet damper

- Recirculation air section, for the white space:

  - Air inlet section
  - Mixing section (bypass), for Cloud Control©
  - Fan section (three fans), including sound attenuators
  - Plate heat exchanger (single, for energy efficiency purposes a double heat exchanger is less beneficial)
  - Filter section (for start-up operations)
  - Cooling coil (chilled water), including chiller
  - Bypass, for Cloud Control©
  - Air outlet damper

Hydraulic Scheme (Figure 5.6)

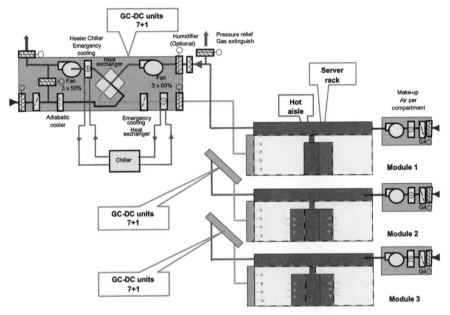

Figure 5.6   System configuration of the GCDC system (DEERNS).

Control

During winter, the system operates in full free-cooling mode. With lower outside temperature, the fan speed is reduced to a minimal required operation. To avoid dehumidification due to the cold air at the heat exchanger, a part of the air is recirculated back on the outside of the heat exchanger. This recirculated air is thoroughly mixed with the outside air to avoid air temperature layers in the supply air. In free-cooling operation, the required cooling is completely supplied with the use of outside air. When the outside temperature becomes higher, the fan speed is increased. With higher outside air temperatures, first the evaporative cooling is put in operation. This will give an amount of additional cooling which depends on air humidity of the outside air. If the evaporative cooling is insufficient, the DX chiller is put into operation, gradually increasing the additional required cooling capacity from the chiller.

Limits of Application

Cooling with outside air depends on the local climate. The maximum extreme temperatures (in summer) and minimum temperatures affect the application of outside aircooling. In winter, below a certain temperature (~6°C), it is necessary to start mixing warm air with outside air to avoid condensation on the plate heat exchangers. In summer, the required temperature of the cooling air for the Data Centre defines the maximum allowable outside temperature. If the outside temperature exceeds this, adiabatic cooling has to be applied or even chilled water cooling (using chillers).

The GCDC © units can operate in partial load, but are limited to minimal airflow required for the heat exchangers to keep a good level of heat transfer. In addition, the fan efficiency decreases in partial load.

The fresh air supply, overpressure and humidity control of the Data Centre have to be done with a separate installation (comparable to computer room air handling units – CRAH units also use a separate fresh air overpressure system).

Economic Aspects

The indirect cooling unit for Data Centres is a relatively new product in the market. There is some cost information available based on a few quotations of a supplier of the GCDC © units.

The cost of GCDC © unit is approximately as follows (2013), see Figure 5.7:

    80 kW: 1,200 €/kW
    120 kW: 1,100 €/kW
    450 kW: 1,000 €/kW

The costs also depend on the required number of units, resulting in a cost cut of the suppliers of these units.

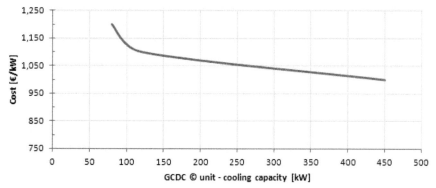

Figure 5.7 Costs of GCDC © unit per kW cooling capacity (DEERNS).

The costs are exclusively for the units, excluding the following:

- Provisions on the building, roof and facade
- Ductwork, air dampers and grilles
- Controls and cabinets
- Electrical provisions

5.2.3 Seawater Air Conditioning System

General Description

Seawater air conditioning systems (SWAC) make use of the cold deep water that can be found within some distance of the coast. At sufficient depth, seawater is not warmed by sunshine nor does heat penetrate there by conduction or convection. This cold water can be used to provide cooling.

The sea server is an inexhaustible source of cold water. The temperature of the water at a particular depth depends on the location. From about 600 m and lower, the seawater temperature reaches what is needed to supply cooling to typical cooling purposes at Data Centres. At 1000 meters and lower, the temperature decline flattens towards the 4°C density peak for water. At very great depths, the pressure causes a shift to even lower temperatures [2]. Generally, the seawater is not applied directly for process cooling. A heat exchanger is used to separate the salty and corrosive seawater from a second loop for Data Centre cooling.

Control

Based on the required cooling water temperature and measured temperature, the seawater pumps are controlled with variable frequency drives. When the temperature of the seawater is lower or equal to the desired chilled water temperature, the chilled water can directly be used for white space cooling by CRAH units.

Seawater cooling can be used through the whole year. An important value is the return water temperature of the seawater. In order to avoid ecological damage to sea life, the outlet temperature must be controlled by mixing the outlet water with cold seawater. The mix ratio is based on the average seawater temperature on outlet depth and the measured temperature of the return water.

Limits of Application

The project needs to be located near the coast. If the location is in a temperate or cold climate, surface seawater can be used, perhaps in combination with a chiller. In that case, the system works similar to aquifer thermal energy storage or cooling with groundwater. If the location is in tropical region, deep-sea water below 600 m is required. Since the cost of the system in this case is determined largely by the piping system in the sea, the limit of application is determined by the distance from the coasts where deep waters are found. This can be within a few hundred meters in some cases, or many kilometres in other cases depending on the location.

Economic Aspects

Costs for a real-world proposed SWAC system have been used to illustrate the economic aspects. The reference system is a 34 MW SWAC with a 220 GWh/year cooling capacity. The seapipe is 8 km out of shore and runs to a depth of 700 m. The seawater is heated by 6 K before being returned to the sea. The electricity cost is 0.26 US$/kWh. Therefore, the cost of sea piping is 33 million US$, the onshore plant (heat exchanger) is 8.6 million US$, electricity cost is 4.6 million US$/year, and operation and maintenance cost is 0.24 million US$. This translates to an investment cost of around 1 $/W for the seawater piping component and 0.25 $/W for the on-shore plant component. Electricity costs and operation and maintenance costs together are 0.02 $ct/kWh. The total investment cost divided simply by 25 years of cooling delivered is 0.008 $ct/kWh.

Location-specific factors contribute greatly to the actual costs. The specific characteristics of the seaside system have a strong influence. Environmental regulations may impose costs on construction of the piping near the shore. The distance to the cold-water resource is another factor. Onshore, issues related to the presence of existing developments can add costs. These all need to be evaluated on a case-by-case basis.

5.2.4 Free Cooling with Groundwater

General Description

Aquifer thermal energy storage (ATES) is based on an open-loop geothermal technology (Figure 5.8). The system stores energy by charging cold groundwater in winter and warm groundwater in summer. The groundwater has to be situated within an aquifer. Therefore, the feasibility for using a warm-cold aquifer/buffer in the earth depends in the soil, which has to be suitable for this type of energy storage. Depending on the feasibility of this, it may be possible to implement one or several aquifer systems.

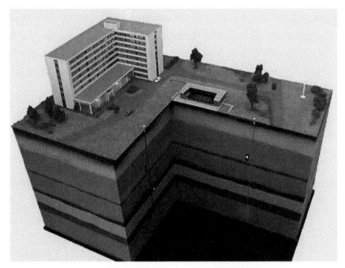

Figure 5.8   Aquifer TES (DEERNS).

Good practice (requirement) of using an ATES system is a balance of storing warm and cold groundwater within the timespan of one year. During the cold season (winter), the groundwater is distributed from the "warm well" to the "cold well" of the aquifer. The "warm" groundwater can be cooled by using free cooling before being fed into the cold well (cold storage). This energy transfer is done using a heat exchanger.

From the cold well, the cold groundwater is used during summer months. After the water has been used for cooling, the warm water is injected into the warm storage of the aquifer. The cycle is repeated seasonally.

ATES consists of (at least) two thermal wells (one cold and one warm well). Other required installation components are the following:

- Pump and filter system
- Heat exchanger in between the ATES system and the cooling and heating installation
- Pipework and valves
- Control systems
- Heat pumps and/or free cooling components (i.e. cooling towers)

The advantage of using an ATES system is the availability of chilled water during the warm season (summer).

Using an ATES system combined with cooling towers gives an advantage in the summer period. Using cooling towers for generating chilled water for the

Data Centre is an energy-efficient way of cooling. However, during summer period, it might not be possible to reach the required chilled water temperature by using (only) cooling towers. In this period, chillers and the ATES system have to "take over" and deliver the required cooling.

Because the ATES system is much more energy efficient, this is used first. As a result, the use of chillers is reduced, because they only are in operation for "peak" cooling in the summer.

Free cooling from, e.g., cooling towers to the ATES is possible during winter period, by minimising the approach of the cooling water in respective to the outside (wet bulb) temperature. Therefore, the cooling towers of a Data Centre can be used for regeneration of the cooling capacity of the aquifer. The design of the cooling towers has to be adjusted to the required injection temperature to the cold well of the aquifer.

The temperature of the cooling water supplied from the cooling towers to the ATES has to be 2–4 K lower than the minimum water temperature in the cold well.

This temperature transition is due to using two heat exchangers:

- Heat exchanger for the cooling towers, to separate water of the cooling tower from the cooling tower for the building installations (i.e. CRAH units)
- Heat exchanger for the ATES system, to separate groundwater from the cooling water system of the Data Centre

Another condition is the cooling capacity of the cooling towers: it has to be equal to the capacity of the ATES.

An essential condition is that the required cooling capacity in summer period may not exceed the regeneration capacity in winter. Otherwise, extra cooling capacity by means of chillers is required in the summer period.

Hydraulic Scheme

The hydraulic scheme of the ATES system (Figure 5.9) consists of the wells, pumps, piping and heat exchangers.

Control

The most important control issue for ATES are the stringent local government rules of the energy balance in the underground wells. During wintertime, the system is charged with cold water from cooling towers or heat pumps (when there is a heat demand, e.g. district heating network). With variable frequency

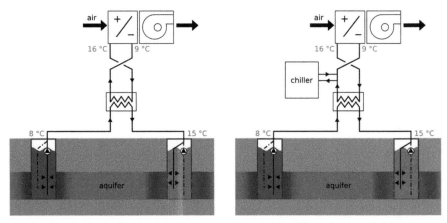

Figure 5.9   Principle of ATES system without or with chiller/heat pump (DEERNS).

drive controlled pumps, the inlet temperature of the cold well is controlled at a minimum of 6°C. Based on the outside temperature, the cooling can be directly delivered by the cooling towers. During summertime with higher outside temperatures, the chilled water for the Data Centre is cooled via a heat exchanger with groundwater from the cold well. Based on the desired chilled water temperature and measured temperature, the well pumps are controlled with variable frequency drives. The chilled water can directly be used for white space cooling by CRAH units.

Limits of Application

Using a warm/cold aquifer in the ground depends on the soil, which has to be suitable for this type of energy storage. Therefore, the local site geology is crucial. If it is possible to use an ATES system, local permits are needed. In addition, the Data Centre has to have a chilled water cooling installation, where the ATES cooling can be incorporated during the summer period.

The required cooling in the summer has to be "regenerated" or "reloaded" in the winter. It is required to have a balance of storing warm and cold groundwater within the timespan of one year.

Economic Aspects

The cost for ATES systems can be divided into two types of ATES: mono-well ATES for smaller capacities ranging from <10 to 50 m³/h (Table 5.2) and doublet-well ATES for larger capacities ranging from 40 to 200 m³/h (Table 5.3). The investment costs for the two types are compared in Figure 5.10.

Table 5.2 Investment cost of a mono-well ATES (costs 2008) [3]

Capacity	Investment Cost	Cost/Flow Rate
10 m³/h	75,000 €	7,500 €/(m³/h)
20 m³/h	125,000 €	6,250 €/(m³/h)
30 m³/h	145,000 €	4,833 €/(m³/h)
40 m³/h	160,000 €	4,000 €/(m³/h)
50 m³/h	170,000 €	3,400 €/(m³/h)

Table 5.3 Investment cost of a doublet-well ATES (cost 2008), based on information from [3]

Capacity	Investment Cost	Cost/Flow Rate
40 m³/h	175,000 €	4,375 €/(m³/h)
80 m³/h	210,000 €	2,625 €/(m³/h)
120 m³/h	255,000 €	2,125 €/(m³/h)
160 m³/h	280,000 €	1,750 €/(m³/h)
200 m³/h	290,000 €	1,450 €/(m³/h)

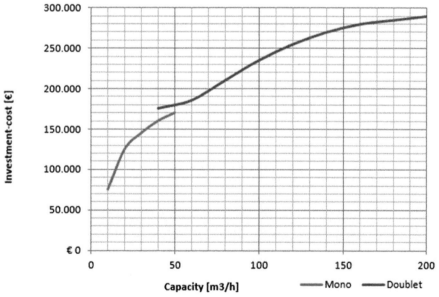

Figure 5.10 Investment costs of a mono- and doublet-well ATES m³/h flow (costs 2008) [3].

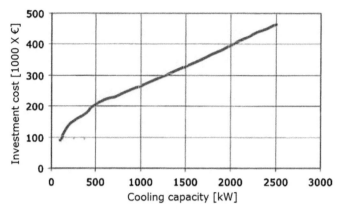

Figure 5.11   Investment costs of an ATES system (2008) per kW cooling capacity with an ATES temperature difference of approximately 6 K [3].

The cooling capacity of the ATES system depends on the temperature difference between warm and cold well. Normally, this is between 6 and 8 K. The maximum cooling capacity of the ATES can be determined by using the ATES design flow and the temperature difference. Based on this, the costs per kW cooling can be calculated (Figure 5.11).

5.3 Increasing Allowable IT Temperatures

Increasing the allowable IT temperatures directly influences the allowable white space temperature. An increased white space temperature leads to improved energy efficiency for any type of cooling concept if proper measures are taken. The degree of effect on the energy efficiency depends on the type of cooling installation which can be air or water cooling.

5.3.1 Increased White Space Temperature with Airside Cooling

Increasing the IT room supply temperature has been suggested as the easiest and most direct way to save energy in Data Centres. However, just implementing a higher inlet air temperature while still relying solely on mechanical cooling[1] may not improve the efficiency of the cooling system. Alternatively, when using air free cooling, raising the maximum allowable white space temperature allows for more annual hours and therefore less hours of additional

---

[1]The term "mechanical cooling" refers to cooling with chiller operation.

cooling by the chiller units. This results in a reduction in the energy demand. In some cases, this may even result in not needing to install chiller units. By raising the maximum allowable white space temperature, evaporative cooling may be sufficient to deliver the required cooling for those "peak" summer conditions. On the other hand, at cold regions, raising the allowable white space temperature may even lead to not needing to apply evaporative cooling anymore.

Moreover, not needing a chiller installation will also affect *PUE* values:

- Lower annual *PUE* results from using more free cooling
- Lower peak *PUE* because no chillers are operating

The lower peak *PUE* also affects the required capacity of the UPS system, which can be reduced. Therefore, the annual energy losses of the UPS system are reduced, leading to an additional increase in energy efficiency.

The impact of raising the maximum allowable white space temperature in reference to the number of annual hours of available free cooling is illustrated in Figure 5.12. The share of free-cooling hours in the total annular number of hours is shown for different locations (Barcelona, Chemnitz, Luleå) and types of airside free cooling (direct, indirect and indirect with evaporative cooling).

The capital expenses costs (*CAPEX*) depend on the location and the increase in allowable white space temperature. Significant cost reduction might be reached when chillers become expendable. Otherwise, implementing free cooling will not cause significant investment cost reduction. Regarding the operational costs (*OPEX*), aspects on the *EER* of the air-cooling units are

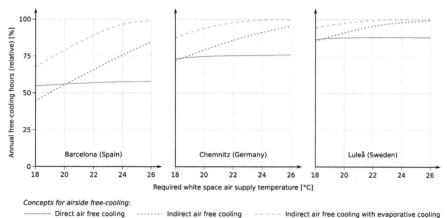

Figure 5.12   Hours of free cooling for different locations and concepts of airside free cooling.

given in Subsections 5.2.1 and 5.2.2. However, a complete analysis of *PUE* and operational costs has to consider, e.g., climate/geographical location and Data Centre allowable temperatures.

5.3.2 Increased White Space Temperature with Chilled Water Cooling

When raising the allowable white space temperature, the energy usage with chilled water cooling systems will significantly decrease due to the following reasons:

- More waterside free cooling can be applied annually.
- The chillers operate more energy efficient.

The increased number of annual hours of waterside free cooling is directly related to raising the allowable white space temperature. The main reason is that the set point of chilled water supply temperature can be raised. This results in an increase in free cooling for all waterside concepts.

Notice that raising the chilled water supply temperature also leads to a decrease in annual energy usage of the chillers. The chillers operate in a more energy efficient way with higher chilled water temperatures (see Figure 5.13).

A point which has to be respected is the maximum allowable chilled water temperature, which depends on the type of chiller. Following ASHRAE and chiller manufacturer recommendations, the maximum chilled water temperature is approximately 16°C. However, the chilled water supply temperature

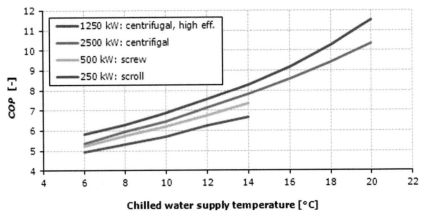

Figure 5.13   Coefficients-of-performance of chillers with condenser water temperature of 30–36°C and variable chilled water temperatures [6].

for the Data Centre can be higher than the allowable chilled water temperature of the chiller. This can be achieved by letting the chilled water partly bypass the chiller. The warm return chilled water is mixed with the cold chilled water from the chillers in order to achieve the required higher chilled water supply temperature.

In conclusion, raising the allowable white space temperature significantly reduces the energy usage for cooling due to more hours of free cooling and more efficient chiller usage.

However, the CRAH units or other type of local cooling equipment do not experience any reduction in energy usage due to the higher allowable Data Centre temperature. Since the required cooling capacity and the temperature difference between air inlet and outlet remain the same, the airflow of the CRAH units and thus their energy demand are constant.

5.3.3 Increasing the Delta *T* Through the IT Equipment

Increasing the Delta *T* (air temperature difference between inlet and outlet) through the IT equipment directly reduces the required airflow rate through the white space. As an example, Figure 5.14 shows the required air volume flow for a 120 kW Data Centre in function of the allowed Delta *T* through the IT equipment. Notice that increasing the Delta *T* by 5 K results in an airflow reduction of more than 30% with the associated energy reduction in the fans.

Moreover, this strategy allows raising the chilled water return temperature. Assuming that the chilled water supply temperature remains the same, the water outlet temperature from the cooling coil (water-air heat exchanger) increases due to a higher air-side Delta *T*. This results in a reduction in the

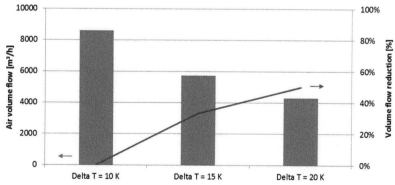

Figure 5.14   Air volume flow required for a standard Data Centre of 120 kW IT power.

required chilled water flow. The benefit of a reduced chilled water flow is that smaller size chilled water pipework can be used. Another option is not to resize the pipework, but use a lower chilled water velocity and lower pressure drop in the chilled water distribution system, which has a positive effect on the required pump energy.

Generally, energy savings depend on the allowable increase in IT temperature. Assuming that the air supply temperature can be increased from 18°C to 24°C, the PUE may decrease by approximately 0.2. This will result in 25% electrical savings for the mechanical installations, which implies a saving of 10% in the total energy demand of the Data Centre. The payback period of this concept will be less than one year, since no extra investments but only modifications in control systems and a different selection of components are required.

5.4 Hot or Cold Aisle Containment

The objective of alternating aisles containments (Figure 5.15) is to separate the source of cooling air from hot air discharge, preventing the cold supply air and hot return air from mixing [9]. Therefore, this strategy can improve predictability and efficiency of traditional cooling systems.

Cold aisles are defined as having perforated floor tiles that allow cooling air to come up from the plenum under the raised floor. The cooling air is distributed to and through the IT racks/equipment and is exhausted from the back of the equipment rack to the adjacent hot aisles. On the other hand, hot aisles do not have perforated tiles. These would mix hot and cold air and thereby lower the temperature of the air returning to the cooling units, which reduces their usable capacity [9]. While hot aisle containment is the preferred solution in all new installations and many retrofit raised floor installations, it may be difficult or expensive to implement due to low headroom or no accessible

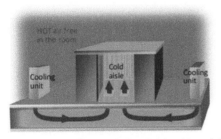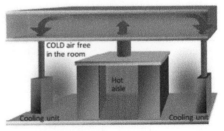

Figure 5.15 Cold and hot aisle containment configuration [7].

dropped ceiling plenum. Cold aisle containment, although not optimal, may be the best feasible option in these cases [7]. In any case, it is essential to separate completely the cold supply air and the warm return air. This way, no temperature mixture takes place and the cooling can be supplied directly to the heat sources (namely the IT racks).

However, it is a common phenomenon with cold and hot aisle containments that the cold air is short-circuiting back to the cooling units (called bypass airflow). This results in only partial usage of the available supplied cooling air. Bypass airflow occurs through unsealed cable cut-out openings and miss-located perforated tiles [10]. Thus, important points related to aisle containments are as follows:

- Blanking panels should be put in unused rack unit positions in equipment cabinets in order to avoid mixing of hot and cold air.
- Unused cabinet/rack positions in equipment rows should be filled with a cabinet/rack or otherwise sealed in order to prevent mixing of air in hot and cold aisles.
- All cable, duct and other penetrations should be airtight.
- Equipment should match the airflow design for the enclosures and white space in which they are placed.

Both hot aisle and cold aisle containment provide significant energy savings compared to traditional uncontained configurations. Niemann et al. (2013) analysed and quantified the energy consumption of both containment methods and concluded that hot aisle containment can provide 43% cooling system energy savings over coldaisle containment mainly due to increased free-cooling hours. They also highlighted that new Data Centre's designs should always use or provision for hot aisle containment.

Generally, aisle containment is estimated to save between 10 and 20% of the cooling system energy, which implies a total saving of 5–10% of the Data Centres total energy demand. Payback period is approximately 5 years for the perforated floor and aisle containments.

5.5 Variable Airflow

The common practice in air-cooled Data Centres is to supply a constant air volume, based on the maximum design-cooling load. This is a robust system without complex control systems. The disadvantage of a constant air volume system is the energy usage: throughout the year, all fans are 100% in operation and require the energy for maximum air recirculation. When the actual heat load is much lower than designed, it may be possible to put a number of units out of operation, but that has to be done "manually" via the building management system (BMS) or on the data floor.

However, the maximum cooling load on which the design is based will never be reached due to the following reasons:

- The data hall/white space may only be partially filled with racks.
- The designed IT power/cooling load is considered as a maximum – in reality, the "maximum" IT rack power will always be lower than designed.
- The IT power might fluctuate during time (depending on the usage of the IT equipment in the racks).

Therefore, the best practice is using variable airflow cooling units. A variable airflow system is based on the actually required cooling, not on the maximum required cooling. Thus, the airflow can be adjusted to the required cooling load. Reducing the airflow implies reduced pressure drop inside the ventilation units. Consequently, running in partial operation gives a significant reduction in the energy usage.

There are different approaches to run a variable airflow strategy: pressure difference, actual IT load and return air temperature.

5.5.1 Strategy A: Pressure Difference

The best and most efficient way for achieving variable airflow is creating a pressure difference between the aisle containment and the data hall where the server racks are situated [5]. In order to achieve this pressure difference (about 10 Pa), the aisle containment has to be airtight. If not, no pressure difference will occur and it will be impossible to control the variable volume system.

In this system, each server controls its required amount of airflow. Each server has one or more integrated fans, which adjust the airflow to the actual power usage of the server. If the power usage of a server decreases, the integrated fans consequently operate on lower speed. This results in a slightly altered pressure difference of the aisle containment and the data hall, leading to reduction in the cooling supply air from the air cooling system (and vice versa).

The advantage of controlling the variable airflow system with pressure difference is that it is performed independently of the temperature difference over the IT racks. This way, each rack can have its own specific air temperature difference, "customised" for that type of server. Therefore, the pressure difference between the aisle containment and the data hall always represents the actual cooling demand.

5.5.2 Strategy B: Actual IT load

The second way to control the variable airflow is by using the actual IT load. This has to be measured per aisle containment. The variable flow is then based on the aisle containment with the highest power density. This system is somewhat less optimal, but also less demanding in regard to the airtightness of the server racks and patch panels. With this method, each aisle receives the airflow that is supplied to the aisle with the highest IT load.

The main disadvantage of controlling the variable flow based on the measured IT load is that this system does not control on the actual required amount of cooling air, but on the average loads and temperatures. Consequently, the supplied cooling air should always be more than the actual required cooling air and there might be a higher risk of hotspots.

5.5.3 Strategy C: Return Air Temperature

A more straightforward, but less efficient way of controlling a variable airflow system is based on the return air temperature to each air-cooling unit. With this control system, the return temperature consists of a mixture of the return temperatures of all different IT racks. However, the maximum allowable return air temperature varies per IT rack. Therefore, the IT rack with the lowest allowable return air temperature determines the maximum allowable return air temperature to the air cooling units. As a result, the supplied cooling air will always be much more than the actually required cooling air and there may be a higher risk of hotspots.

As mentioned before, a variable airflow system is based on the actually required cooling, not on the maximum required cooling. Thus, the airflow can be adjusted to the required cooling load leading to reduce pressure drops in the ventilation units. Additionally, a higher air Delta T is reached which implicates the possibility for higher chilled water Delta T and thus reduced water volume flow rates as well. Consequently, running in partial operation gives a significant reduction in energy usage. However, an important point with variable flow air cooling units is that the partial load is limited since CRAH units have a minimum partial load of approximately 40%. Lower airflows will result in insufficient pressure and cooling capacity.

The energy usage of fans and pumps[2] decreases significantly when their speed is reduced. For the case of a fan, the required power depends on the airflow, the pressure drop and the fan efficiency. It can be calculated as follows:

$$P_{\text{fan}} = \frac{\dot{V} \cdot \Delta p}{\eta_{\text{fan}}} \tag{5.3}$$

where

P_{fan} = fan power consumption [W]
$\dot{V}$ = airflow [m³/s]
Δp = pressure drop [Pa]
η_{fan} = fan efficiency [-]

A reduction in the required airflow results in a reduction in the required fan power as both $\dot{V}$ and Δp are reduced in the equation. Since the airflow has a linear relation to the air speed and the pressure drop has a 2nd power relation to the air speed, theoretically, the energy demand of the fan decreases by the 3rd power. This relation is shown by the following equation:

$$\frac{P_{\text{fan},1}}{P_{\text{fan},2}} \sim \frac{\Delta p_1^2}{\Delta p_2^2} \sim \frac{\dot{V}_1^3}{\dot{V}_2^3} \tag{5.4}$$

However, the efficiency is not constant but varies in function of the fan load. It has a maximum value at the ideal working point and decreases in any other situation. Thus, a dynamic study for each case has to be done, taking into account the working parameters of the specific fan. Therefore, at high-level design with variable flows from 50 to 100%, the reduction in the required fan power is approximately to the power of 2.5 instead of the 3rd power.

Figure 5.16 shows an example of fan power consumption in function of its volume flow ratio. For example, when the velocity of the fan is reduced by 25%, fan power consumption decreases to about 50%. Please notice that these numbers are just given to understand the phenomenon and should be recalculated for specific fans.

As far as pumps are concerned, it has to be noticed that the reduction in the pressure drop depends on the hydraulic system of the pump. This may result inless energy reduction, as control valves may require a constant pressure drop. In that case, this has to be corrected in regard to the energy reduction.

---

[2]The flow through evaporator and condenser of a chiller may not be reduced too much in order to prevent the heat transfer from dropping. Reduced water flow is especially interesting for the chilled water distribution circuit.

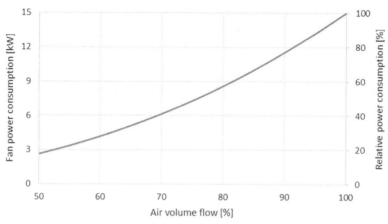

Figure 5.16 Power consumption of a 15 kW fan in function of the airflow ratio.

Assuming the average IT load of a Data Centre to be 50–75% of the design load, applying variable flow will result in approximately 15–20% electrical savings for total mechanical installations, which implies a saving of 8–10% of the total energy demand. The payback period is estimated to be approximately 5–7 years for control systems and airtight containments.

5.6 Partial Load – Redundant or Oversized Components

When using components in partial load, the energy usage will decrease. This applies to all components such as fans, pumps and chillers. Moreover, an increase in energy efficiency can also be achieved. However, this depends completely on the specifications of the different components and on the percentage of partial load (approximately 70%). Especially with chillers, partial load has positive effects on the efficiency. In this section, first it is described how to achieve partial load operation and why it might be beneficial to use redundant and/or oversized components. Then, the effect of partial loads for different mechanical components is discussed.

5.6.1 Redundant Components and Oversizing Components

The components of the cooling installation in a Data Centre can operate in partial load due to the following reasons:

- The actual IT power is usually (much) less than the contracted/designed IT power.

- Using redundant components: Most Data Centre installations consist of redundant components (for service/maintenance and fail-safe purposes). When all components are used – including the redundant components – the installation runs in partial load (e.g. with an $N + 1$ configuration, when N = 4, all components can operate at 80% of their full load).
- Oversizing components for more efficiency: Oversizing the capacity of components makes it possible to run on partial load, even when no redundant units are available (Data Centre with N-configuration).

The effect of the partial load on the energy efficiency depends on the installation component. Therefore, the effect of partial load on common used components for cooling installations of Data Centres is discussed first for chillers and then for fans and pumps.

5.6.2 Partial Load with Chillers

It is well known that most chillers operate more efficiently in partial load using variable speed compressors and pumps, resulting in a reduced (variable) water and refrigerant flow. This is illustrated with four types of chillers: high-efficiency centrifugal compressors (Figure 5.17), centrifugal compressors (Figure 5.18), screw compressors (Figure 5.19) and scroll compressors (Figure 5.20). For each type of chiller, a full load and 50% partial load curve is shown. For each chiller, the 50% load operation is more efficient than the operation at 100% load[3].

Notice that the characteristics of partial load differ per type and design of the chillers, so these are just examples to illustrate the benefit of partial load on the energy efficiency. In addition, the performance of the chillers with partial load is limited. Minimal partial loads are, e.g., approximately 15–25% depending on the type of chiller. Concerning the energy efficiency, an optimum exists at a certain partial load. Below this, the energy efficiency starts to decrease again.

5.6.3 Variable Flow with Fans and Pumps

As mentioned before, the energy usage of fans and pumps decreases significantly when operating in partial load. For variable flows between 50 and

---

[3]Cooling water temperature was 30–36°C for 100% and 30–33°C for 50% (Johnson Control 2014). However the auxiliary energy has to be taken into account. For wet cooling towers 27°C is assumed.

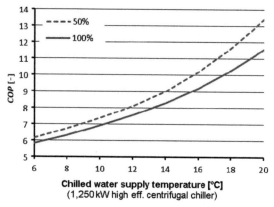

Figure 5.17   *COP* of a high-efficiency centrifugal chiller at 50% and 100% cooling load [6].

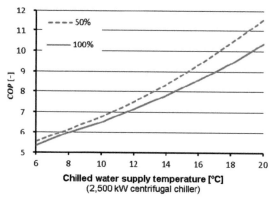

Figure 5.18   *COP* of a centrifugal chiller with at 50% and 100% cooling load [6].

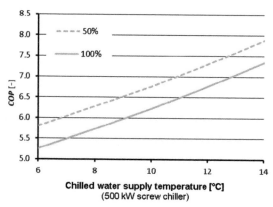

Figure 5.19   *COP* of a screw compressor chiller at 50% and 100% cooling load [6].

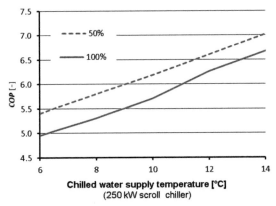

Figure 5.20   *COP* of a scroll compressor chiller at 50% and 100% cooling load [6].

100%, the power demand of fans and pumps approximately decreases by the power of 2.5 with reduced flow. However, splitting a constant water or airflow between several pumps or fans (redundant components) only reduces the total energy demand if the efficiency of each component is higher for the reduced flow than for design flow.

5.6.4 Oversizing Dry Coolers and Cooling Towers

Dry coolers or wet cooling towers are commonly used for rejection of waste heat from water-cooled chillers. An oversized dry cooler or cooling tower is able to generate free cooling at higher outside air temperatures. This is because the approach between the outside temperature and the cooling water supply temperature from the cooler decreases with an oversized cooler. Thus, the annual amount of free cooling increases when using oversized coolers.

5.6.5 Energy Savings and Payback Periods

Generally, energy demand reductions of 20–40% will be achievable for most components when operating in partial load. This results in a total electrical saving of 10–20% for the mechanical installations, which implies a saving of 6–12% related to the Data Centre's total energy demand. Payback period is less than one year when redundant components are used since that does not require extra investments. Oversizing components leads to extra investments; in this case, the typical payback period will be 7–10 years.

5.7 High Energy Efficiency Components

5.7.1 Fans and Pumps

Energy efficiency can be increased by selecting components that are specifically designed for the cooling specifications of the white space. Examples of these are distribution components such as follows:

- High-efficiency, direct-driven and variable speed fans for the CRAH units
- High-efficiency, variable speed pumps

Therefore, the energy efficiency of these components should not only be assessed at 100% load, but also on partial loads (i.e. 25%, 50% and 75% load). The improvement of the energy efficiency may lead to energy reductions for fans and pumps varying from 10 to 30%.

5.7.2 Air-Cooled Chillers

A free cooling option can be incorporated in an air-cooled chiller, although this is very limited. Only at very low outside temperatures, it is possible to generate the required chilled water without using compression cooling. This can be optimised by using advanced high-tech free-cooling chillers. These types of chillers have an optimised free-cooling module using bypass piping, a dedicated pump for free cooling and a high-efficient heat exchanger. In addition, high-efficient compressors (with magnetic bearings, using variable speed permanent magnet) are used. This results in an improvement of the energy efficiency of the chiller:

- Significantly more annual hours of free cooling are available. Normally, the temperature difference between the outside temperature and the chilled water temperature is 8–10 K, while with the high-efficiency chiller, this can be reduced to 2–3 K according to quotations and technical specifications from APAC air conditioning.
- At free-cooling mode, the *EER* will be approximately twice as high (50% reduction in energy usage) compared at the same outside temperatures.
- At "standard" compression cooling operation, the efficiency is also higher (approximately 25% higher *EER*).

However, the cost of these high-tech free-cooling chillers is significantly higher than the "standard" air-cooled chillers. As an example, chillers with a cooling capacity of 250 kW and a chilled water supply temperature of 16°C are compared[4]:

---

[4]Based on quotations from Carrier and Apac Air-conditioning.

- Standard free cooling air-cooled chiller ~150 €/kW
- High-efficiency free-cooling air-cooled chiller ~250 €/kW

Therefore, the economic feasibility has to be studied before implementing such systems in the infrastructure.

5.7.3 Water-Cooled Chillers

Water-cooled chillers are available as high-efficiency performance chillers as well. This is also achieved by using magnetic bearing technology with variable speed drives. An example of this is a comparison of a standard and a highly efficient chiller, as is shown in Figure 5.21. The improvement of the efficiency of the high-efficiency chiller depends on various items, such as the following:

- Condenser water inlet temperature
- Chilled water inlet temperature
- Partial load

This results in an *EER* being 5–50% higher than with the standard centrifugal chiller. However, the costs of the high-tech centrifugal chillers are also significantly higher than the standard chillers. The estimated costs[5] of different water-cooled chillers are as follows:

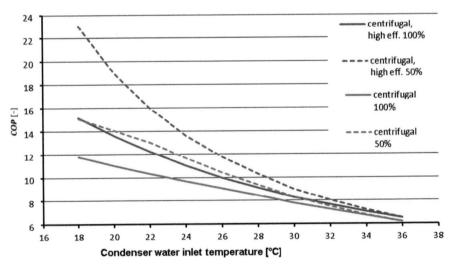

Figure 5.21 Comparison of a "standard" and a high-efficiency centrifugal chiller with variable condenser water inlet temperatures [6].

---

[5]Based on quotations from different suppliers, mainly from York (Johnson Control).

- Scrollcompressor, cooling approximately 250 kW ∼125 €/kW
- Screwcompressor, cooling approximately 500 kW ∼120 €/kW
- Centrifugalcompressor, cooling approximately 2500 kW ∼100 €/kW
- High-efficiency centrifugal compressor, cooling approximately 1250 kW ∼125 €/kW

Therefore, as in the previous subchapter, the economic feasibility has to be studied before the implementation of such systems in the infrastructure.

Generally, the impact of using highly efficient components depends on the type of component. Normally, an energy saving of 15–25% can be achieved for a number of components. This has a total impact on the electrical savings for the mechanical installations of approximately 10% leading to savings in the total energy demand of about 5%. High-energy-efficiency components will lead to extra investments with payback periods being 10 years or less.

5.8 Conclusions

Data Centres not only need electrical energy to run the IT equipment but also need a lot of energy for cooling, i.e. for removing the heat generated by the IT hardware. Six advanced technical concepts are proposed in this chapter which lead to a reduction in the electric energy required for cooling. Free cooling and increased allowable IT temperature additionally reduce the load of mechanical cooling. With free cooling, heat is transferred actively to the environment without using a mechanical chiller, while increased IT temperatures lead to better operational requirements of the equipment (chiller) and also enhance the hours of free cooling.

Further concepts, hot/cold aisle containment and variable airflowcover the optimisation of air and chilled water flows by preventing mixing of hot and cold air and adapting the volume flow to the load. This leads to increasing return temperatures and temperature differences, which are favourable both for electricity demand and for potential heat reuse. Concepts, i.e. at partial load with redundant or oversized components and highly efficient components, deal with increasing the efficiency of components as chillers, fans and pumps by running them under part-load conditions and by implementing highly efficient products.

The next step after load minimisation by means of efficiency measures is to supply the electrical and cooling load efficiently and with a high share of renewable energy resources. This subject will be discussed in the next Chapter 6.

References

[1] APC by Schneider Electric. (2014). *Cooling economizer mode PUE calculator*. Retrieved from http://tools.apc.com

[2] Bergman, J. (2011, February 11). *Windows to the universe. Retrieved* from http://www.windows2universe.org/earth/Water/temp.html

[3] Bodemenergie, N. L. (2014). Retrieved from www.bodemenergienl.nl

[4] DEERNS. (n.d.). Retrieved January 2015, from https://www.deerns.com/

[5] Hundertmark, T. (2012). *Patent No. EP2596295A2, US20130176675, WO2012011813A3*. World Intellectual Property.

[6] Johnson Control. (2014). York Chillers. The Netherlands.

[7] Newman, J., Brown, K., and Avelar, V. (2013). *Impact of hot and cold aisle containment on data center temperature and efficiency*. Schneider Electric – Data Center Science Center.

[8] OTEC Corporation. (n.d.).

[9] Sullivan, R. F. (2002). *Alternating cold and hot aisles provides more reliable cooling for server farms*. Santa Fe: The Uptime Institute, Inc.

[10] Sullivan, R., Strong, L., & Brill, K. (2004, 2006). *Reducing Bypass Airflow is Essential for Eliminating Computer Room Hot Spots*. Santa Fe: The Uptime Institute, Inc.

6

Advanced Technical Concepts for Power and Cooling Supply with Renewables

Verena Rudolf[1], Nirendra Lal Shrestha[1], Noah Pflugradt[1], Eduard Oró[2], Thorsten Urbaneck[1] and Jaume Salom[2]

[1]Chemnitz University of Technology, Professorship Technical Thermodynamics, Germany
[2]Catalonia Institute for Energy Research – IREC, Spain

6.1 Introduction

The next step after load minimisation by means of energy efficiency measures as described in Chapters 4 and 5 is to supply the electrical and cooling load efficiently with a high share of renewable energy resources. Six advanced technical concepts containing promising energy-efficiency strategies are proposed in this chapter, which have the potential to minimise the non-renewable primary energy demand for supplying both cooling and power or only cooling to Data Centres.

At first, four scenarios of Data Centres defining sizes and redundancy levels are presented. Advanced technical concepts are assigned to these scenarios in order to cover a wide range of Data Centre types. A brief description of each concept including thermal[1] and electric schemes along with the main components is presented. As far as thermal and electrical storages are concerned, operation and management strategies are required for optimal integration into Data Centres and in order to reach high-energy efficiency. These strategies are incorporated in the concepts as well. Energy flows per year using a Sankey chart are presented for selected scenarios with the objective to present how energy is distributed within the different subsystems.

---

[1]The schemes are simplified and do not consider safety equipment and other parts which are required in practice.

6.1.1 Concepts Overview

The advanced technical concepts are defined as a combination of known technologies that are adapted to the specific requirements of Data Centres. It is derived from the integration of innovations under a holistic approach with the aims to provide power, cooling, or both for a Data Centre and to ensure a high share of renewables. Thus, the integrated solutions of the concepts are evaluated and not the sum of the individual subsystems.

The concepts have been chosen based on the following premises:

- The concepts should cover the whole size range of Data Centres.
- The concepts should cover the variety of geographical and climatic conditions in Europe as well.
- The concepts should have the potential to reach a high share of renewables and efficient supply of Data Centres in urban locations.
- The concepts should be economically feasible at present or supposed to become feasible within the next years.

As the most promising solutions, six thermal and electrical power supply concepts for Data Centres have been chosen to be presented in this chapter (Table 6.1). Three of the proposed concepts (1, 5 and 6) cover both cooling and power supply including power generation technologies. These could supply the cooling load as well as a high share of the total electric energy demand of the Data Centre (if necessary, additional electricity might be purchased from the grid). Concept 4 is based on electricity purchased from the grid, but includes thermal and also electric storages for optimal utilisation of the grid power (for example, the storage can be charged when the current electricity price is low or when a lot of renewable energy is fed into the grid leading to a high share of renewables). The other two concepts 2 and 3 are pure thermal concepts, which do not include power supply. Purchasing green electricity from the grid is recommended when implementing these concepts in order to minimise the Data Centre's non-renewable primary energy demand.

Four locations, Barcelona (Spain), Frankfurt (Germany), Amsterdam (Netherlands) and Stockholm (Sweden), representing different climate conditions around Europe, were chosen as examples for evaluating the concepts. It is important to keep in mind that these locations do not represent the climate conditions of the entire country, but are an example for the conditions in the climate zone they are located. The assumed location determines the number of annual hours which allow free cooling, the availability of renewable resources such as solar radiation and wind, the share of renewable energy in the national

Table 6.1 Characteristics of energy supply systems for Data Centres [39]

No. of Concept/Scheme		1	2	3	4	5	6
Liquid-cooled server			x				
Free Cooling	Air	x	x	x	x	x	x
	Cooling Tower			x			x
Chiller	Compression	x		x	x		
	Absorption					x	x
Cold Storage	Buffer, Cold Water	x		x		x	x
	Cold Water				x		
District Cooling			x				
National Power Grid		x	x	x	x	x	x
Battery Storage	Lead Acid	x					
	Lithium Ion				x		
Renewable Energy sources for Power and Possible	PV	x					
	Wind	x					
Heat Production	CHP with Biomass					FC	CHP
Heat Storage (Buffer, Warm Water)						x	x
Heat Feed-In			x			x	x
Heat Pump for Temperature Rise			x				

grid and the electricity price. Free cooling has been considered in all supply concepts, because it makes sense from the point of view of the primary energy demand at least during winter in all locations.

6.1.1.1 Sankey charts analysis

Within the concept description, the simulation results in terms of Sankey charts illustrate the distribution of average energy flows per year within different subsystems for different scenarios (Table 6.2).

The dynamic simulation tool TRNSYS 17 is used for modelling the six concepts. Each model consists of a group of macros that are connected and create an energy model. The workload has been assumed as the HPC workload (constant over time) for each concept during the simulation. The design IT power of the systems ranges from 120 kW to 2000 kW. The actual IT power consumption of the system is smaller than the design IT power. In general, the IT power consumption of the Data Centre depends on many factors, such as the server type, physical properties of the cooling medium (such as the temperature, flow rate), workload and the occupancy rate. Therefore, in reality, the IT load of the system can never reach the value of the design IT power. This is also regulated by the safety margin.

6.2 Description of the Proposed Advanced Technical Concepts

In this chapter, a brief description of the aforementioned concepts including thermal[2] and electric schemes along with the main components is presented. Furthermore, operation and control of the concepts as well as their limits of

Table 6.2   Data Centre scenarios

	Scenario 1	Scenario 2	Scenario 3	Scenario 4
Type of facility (example)	Server room	Very small enterprise Data Centre	Small enterprise Data Centre	Cloud Data Centre
Localisation	In house	In house	Building for the DC in a build environment	
Assumptions for concept development	24 kW Redundancy level I	120 kW Redundancy level II	400 kW Redundancy level III	2000 kW Redundancy level III

---

[2]More detailed hydraulic schemes of the subsystems as well as information about their control, requirements and costs can be found in annex A. The schemes are simplified and do not consider safety equipment and other parts which are required in practice.

application and geographic restrictions are described. Simulation results in terms of Sankey diagrams illustrate the specific energy flow of each concept.

Emphasis is placed on the thermal systems (for cooling supply) because they are more complex than the power supply system due to energy conversion and heat transfer processes. Furthermore, the grid is always available for purchasing additional electricity or feeding excess power from own generation, whereas the cooling load has to be supplied just in time by own generation or storage discharging. However, the importance of an optimised and efficient power supply system should not be neglected.

In order to guarantee the reliable operation of Data Centres, redundancy and backup equipment are considered in the concepts according to the assumed redundancy levels of the scenarios.

6.2.1 Photovoltaic System and Wind Turbines with Vapour-Compression Chiller and Lead-Acid Batteries

General

In concept 1, vapour-compression chillers along with a dry cooler are used to produce cooling energy during summer. Figures 6.1, 6.2 and 6.3 depict the thermal and electric scheme of this concept. The electrical power required to drive the chiller and to run the IT hardware can be generated by a photovoltaic system and wind turbines installed near the building; additional power is purchased from the grid. Lead-acid batteries are used for decoupling power generation from power consumption and cooling demand. Thus, batteries are charged for example when renewable electric generation is high or when the cost of electricity is low. This strategy allows adopting the Data Centre's

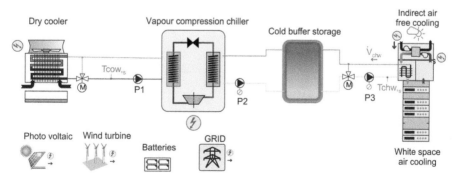

Figure 6.1 Thermal scheme of concept 1.

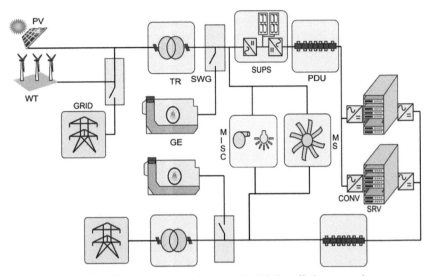

Figure 6.2   Electric scheme of concept 1 with the off-site generation.

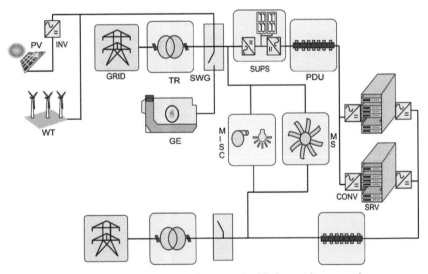

Figure 6.3   Electric scheme of concept 1 with the on-site generation.

total energy draft from the grid to the fluctuating parameters (e.g., cost and share of renewables) in order to optimise the Data Centre energy supply. Additionally, time shifting of the IT workload should be applied in this concept (see Chapter 4).

In winter, indirect air free cooling is performed for efficient cooling supply to the Data Centre.

Operation and Control

Figure 6.4 depicts the control strategy of concept 1. The cooling control strategies vary depending on the operating parameters such as the ambient air temperature, wind and solar power availability, share of renewable power in the grid, cost of electricity and the state of charge of the battery. The operating mode of the concept is selected based on the availability of one of the following order of the operating parameters; first, the availability of the wind and solar power, secondly high share of renewable power in the grid and third cheap electricity. In some scenario, there is a possibility of the availability of more than one operating parameter. The set point values for the supply temperatures of the chilled water ($T_{chw,s}$) and the cooling water ($T_{cow,s}$) used in this concept are 10°C and 27°C, respectively. The pump used for the cooling water circuit (i.e. P1) could be operated with constant speed, whereas the pumps (i.e. P2 and P3) could be operated with variable speed. The flow rate of fluid through the chiller water pump P3 ($\dot{V}_{chw}$) is controlled in order to maintain a specific supply air temperature of 20°C into the white space.

For efficient part load operation of the cooling system, chillers as well as cooling tower units should be sequenced.

Limits of Application

This concept is suitable for all kinds of Data Centres. However, this expensive combined solution requires relatively high levels of radiation and high availability of wind resources to justify the extra investment which is needed. Additionally, sufficient space is required for both the PV panels and the wind turbines.

Backup and Redundancy

As a redundancy, $N + 1$ compression chillers and wet cooling towers are used to attain redundancy level III. All components are connected by two independent paths (Table 6.3).

Sankey Analysis

For the simulation of the concept, a 400 kW IT power capacity Data Centre located in Barcelona has been used. Moreover, the parameters such as IT power

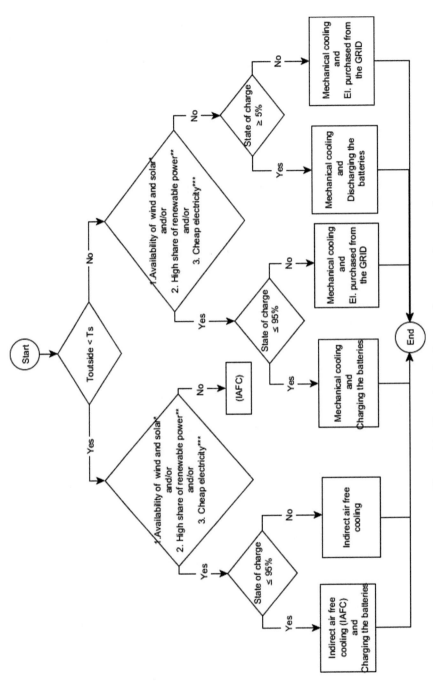

Figure 6.4   Flow chart for the control strategy of concept 1.

Table 6.3 Subsystems in concept 1

Subsystem	Comments
PV power system	Grid-tie photovoltaic power system installed near the building
Wind turbine system	Grid-tie wind turbines installed near the building
Lead-acid battery	Sized according to the required electrical power for IT and cooling distribution as well as the envisaged charging and discharging time
Vapour-compression chiller	Highly efficient machine, e.g., with screw compressor
Dry cooler	Sized according to the chiller capacity
Cold buffer storage	Allowing for optimal operation of both the cold generator (compression chiller) and the cold consumer (e.g., CRAH) by smoothing temperature fluctuations
Indirect air free cooling	Could include adiabatic cooling; run as often as possible

capacity, location and the redundancy level from the scenario 3 (Table 6.2) are applied. The yearly average value of the share of renewable energy in the national grid (RES) is 0.36 during 2013 [16, 17]. Figure 6.5 depicts

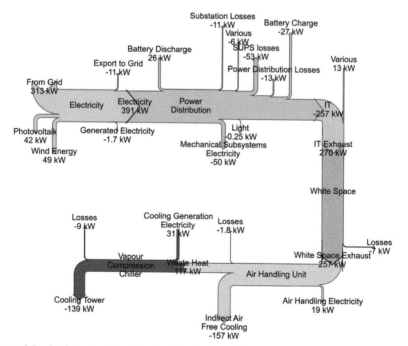

Figure 6.5 Sankey chart showing the distribution of average energy flows per year within different subsystems of concept 1 for scenario 3 (Boundary condition: Barcelona, 400 kW IT power capacity, RES = 0.36).

the distribution of the average energy flows per year within the different subsystems of concept 1. The electricity generated by the PV system and the wind turbines is not sufficient and therefore purchased from the national grid. In the case when the electricity generated is not self-consumed, it is exported to the grid. The concept uses a battery, and about 7% of the electricity is stored temporarily in the battery. It is visible that the free cooling covers only 57% of the cooling load of the Data Centre and the rest is supplied by the chiller.

6.2.2 District Cooling and Heat Reuse

General

In concept 2, chilled water for air cooling is supplied by a district cooling system. Additionally, heat from direct liquid cooling is reused for space heating by means of a heat pump.

Figures 6.6 and 6.7 show the thermal and electric schemes of this concept, respectively. During summer, chilled water from the district cooling system is used to cool the air flowing into the Data Centre, and during winter, indirect free air cooling is conducted. The water for direct liquid cooling is cooled by a heat pump, which provides heat for space heating and domestic hot water. A dry cooler could be used if there is no heat demand.

A district cooling water network with a high share of cooling from renewable sources is required for implementing this concept, but there are no geographical restrictions. The concept might be applied to very small as well as very large Data Centres (50 kW to >10 MW) if a suitable heat sink for heat reuse is available. As the concept does not cover power generation, the required electrical energy can be purchased from the national grid for instance.

Main Components

The main components for this concept are listed in Table 6.4.

Operation and Control

Figure 6.8 depicts the cooling control strategy for both the air-cooled and direct liquid-cooling mode of concept 2. The chilled water volume flow on the primary side of HEX is controlled according to the air-cooling demand[3],

---

[3]The flow rate of fluid through the pump P1 ($\dot{V}_{chw}$, 1) is controlled in order to maintain a specific supply air temperature of 20°C into the white space.

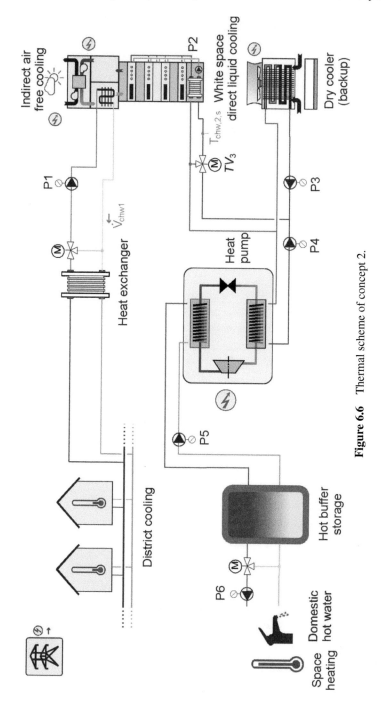

Figure 6.6 Thermal scheme of concept 2.

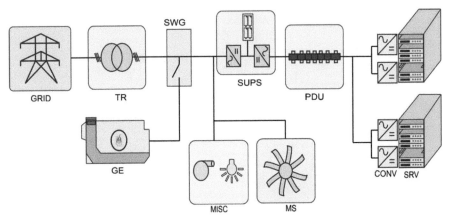

Figure 6.7 Electric scheme of concept 2.

Table 6.4 Subsystems in concept 2

Subsystem	Comments
District cooling system	Supplies water with temperature of, e.g., 5/10°C (higher temperature is preferable for efficiency reasons, e.g., 10/16°C)
Heat exchanger	Water/water plate heat exchanger; sized according to the mean temperature difference and required air cooling power
Heat pump	CO_2 high-temperature heat pump; sized according to water cooling power demand; might not be necessary in a Low-Ex heating system without domestic hot water production
Hot buffer storage	Allowing for optimal operation of both the heat generator (heat pump) and the heat consumer (domestic hot water and space heating) by smoothing temperature fluctuations; works as hydraulic separator between generator and consumer circuit
Space heating system	Design supply temperature as low as possible, e.g., Low-Ex system (40/35°C)
Indirect air free cooling	Could include adiabatic cooling; run as often as possible

while the heat pump is operated depending on the liquid cooling demand. Both the hot and the cold water supply temperatures of the heat pump have to be maintained in a given range. A set point value is defined for one of the temperatures, while the other one is monitored. In concept 2, temperature of the cold water supply (i.e. $T_{chw,2,s}$) is set as 40°C. All the pumps used in concept 2 could be operated with the variable speed.

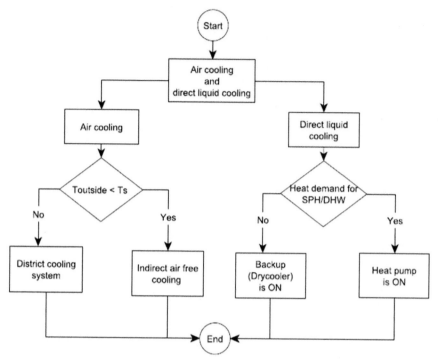

Figure 6.8 Flow chart for the cooling control strategy of concept 2.

Limits of Application

The availability of a district cooling system limits the use of the proposed concept (only urban location). Furthermore, there must be an appropriate heat demand for reusing the heat from direct liquid cooling.

Backup and Redundancy

Redundancy level II can be attained by installing dry cooling tower units, which dissipate the heat from direct liquid cooling when the heat pump is not operating. Alternatively, the direct liquid cooling circuit could be connected to the district cooling system as well. For this system, a guarantee of >99% availability given by the provider is assumed and no backup is considered at the Data Centre.

Sankey Analysis

For the simulation, a 120 kW IT power capacity Data Centre located in Frankfurt and the parameters from scenario 2 (Table 6.2) are used. Figure 6.9

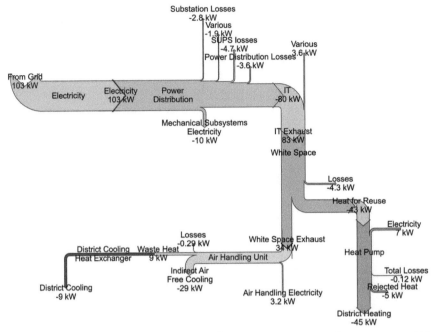

Figure 6.9   Sankey chart showing the distribution of average energy flows per year within different subsystems of concept 2 for scenario 2 (Boundary condition: Frankfurt, 120 kW IT power capacity, RES = 0.26).

illustrates the distribution of the average energy flows per year within the different subsystems of concept 2. Here, the Data Centre is cooled by the air-cooling and the liquid-cooling system. In order to cool the air of the Data Centre in the summer, chilled water from the district cooling system is used, and during the winter, indirect air free cooling is applied. The air-cooling system accounts for 74%, while the liquid-cooling system accounts for 24% of the total cooling load of the Data Centre.

6.2.3 Grid-Fed Wet Cooling Tower Without Chiller

General

In concept 3 (Figure 6.10), wet cooling towers (without chillers) can be used to produce cooling energy. When this evaporative free cooling is not possible, backup vapour-compression chillers along with cooling towers are used. The electrical power required to drive the cooling towers and the backup chillers can be purchased from the national grid. In winter, direct air free cooling

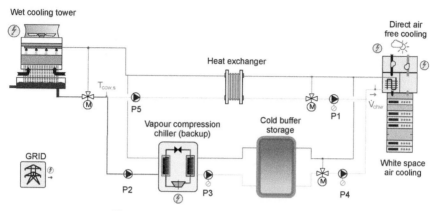

Figure 6.10 Thermal scheme of concept 3.

is performed for efficient cooling supply to the Data Centre. Figure 6.11 depicts the electric scheme of this concept.

This concept might be applied in very small as well as medium-sized Data Centres (50 kW to 1 MW). Table 6.5 gives an overview on the main components (subsystems) shown in Figure 6.10.

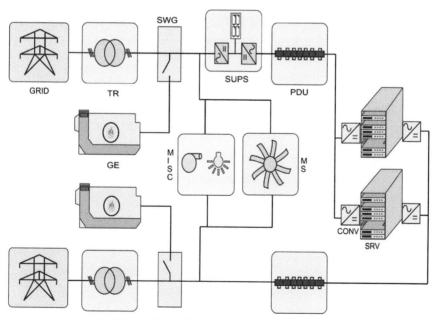

Figure 6.11 Electric scheme of concept 3.

Table 6.5   Subsystems in concept 3

Subsystem	Comments
Wet cooling tower	Used for free cooling and for dissipating the heat from the chiller in summer
Vapour-compression chiller	Backup vapour compression chiller; highly efficient machine, e.g., with screw compressor
Heat exchanger	Water/water plate heat exchanger; sized according to the mean temperature difference and required air cooling power
Cold buffer storage	Allowing for optimal operation both of the cold generator (backup compression chiller) and the cold consumer (e.g., CRAH or cooling coil) by smoothing temperature fluctuations
Direct air free cooling	Run as often as possible

Operation and Control

Figure 6.12 shows the cooling control strategy of concept 3. In case that the outside air temperature is lower than the supplied air temperature to the IT room, the direct air free cooling is used. In other cases when the outside air temperature is higher than the supplied air temperature, the direct air free cooling is not applicable. In this scenario, the cooling tower is used to produce the cooling energy. The operation of the cooling tower control is based on the ambient wet-bulb temperature. This evaporative free cooling is only possible when the outside wet-bulb temperature is lower than the limit set point.[4] In the case when the outside wet-bulb temperature is higher, backup vapour-compression chillers along with dry cooling towers are used. The set point value for the supply temperature of the cooling water ($T_{cow,s}$) used in this concept is 10°C. The pump used for the cooling water circuit of the backup chiller (i.e. P2) is the constant speed pump, and the pumps (i.e. P2, P3, and P4) are variable speed pumps. The flow rate of fluid through the cooling water pump P1 ($\dot{V}_{chw}$) is controlled in order to maintain a specific supply air temperature of 20°C into the white space.

If the implementation of the cooling system has multiple cooling towers and chillers, efficient part load operation sequenced operation strategies should be used.

Limits of Application

The feasibility of this concept depends largely on the number of free-cooling hours and the possibility of adiabatic cooling by direct use of the cooling tower.

---

[4]The limit value for the allowable ambient wet-bulb temperature is 6°C.

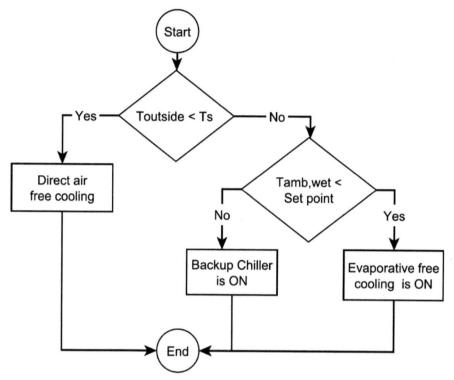

Figure 6.12 Flow chart for the cooling control strategy of concept 3.

For this reason, the maximum wet-bulb temperature at the location must not be too high for providing the chilled water temperature required in the Data Centre.

Plenty of makeup water is required for the application of wet cooling towers. Furthermore, for the application and maintenance of cooling towers and their water circuit in it, certain health and safety regulations need to be fulfilled in some countries. These local regulations may restrict the application of wet cooling towers.

Backup and Redundancy

This evaporative free cooling system needs a backup system when the environmental conditions are not suitable. As a redundancy, $N + 1$ compression chillers and wet cooling towers are used to attain redundancy level III. All components are connected by two independent paths.

Sankey Analysis

Figure 6.13 depicts the distribution of the average energy flows per year within the different subsystems of concept 3. For the simulation, a 400 kW IT power capacity Data Centre located in Stockholm and the parameters from the scenario 3 are used. The simulation results show that 96% of the cooling load of the Data Centre is covered by the direct air free cooling and the wet cooling tower is operated to meet the remaining cooling load.

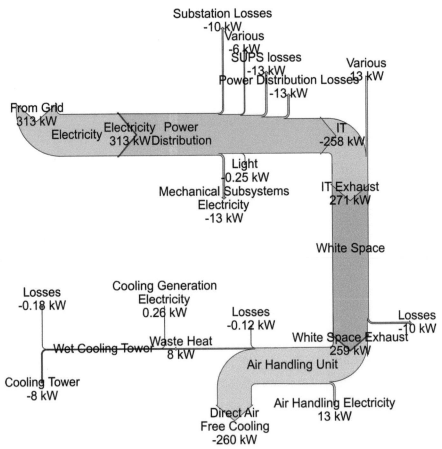

Figure 6.13   Sankey chart showing the distribution of average energy flows per year within different subsystems of concept 3 for scenario 3 (Boundary condition: Stockholm, 400 kW IT power capacity, RES = 0.62).

6.2.4 Grid-Fed Vapour-Compression Chiller with Electrical Energy and Chilled Water Storages

General

In concept 4 (Figure 6.14), vapour-compression chillers along with wet cooling towers are used to produce cooling energy during summer. The electrical power required to drive the chiller can be purchased from the national grid. A large chilled water storage tank (CHWST) for decoupling cooling generation from cooling demand and lithium-ion batteries for storing electrical energy are provided. Thus, both storages are charged, for example when the cost of electricity is low or when the share of renewables is high in the grid. This strategy allows adopting Data Centre's total energy draft from the grid to the fluctuating parameters (e.g., cost and share of renewables) in order to optimise the Data Centre energy supply. Additionally, charging the storage during the colder night might be advantageous especially in warmer regions because cooling tower operation requires less energy when the ambient temperature is lower. During winter, indirect air free cooling is performed for efficient cooling supply to the Data Centre. The concept might be applied in small as well as large Data Centres (50 kW to 10 MW). Figure 6.15 depicts the electric scheme of this concept.

Table 6.6 shows an overview of the main components of the concept.

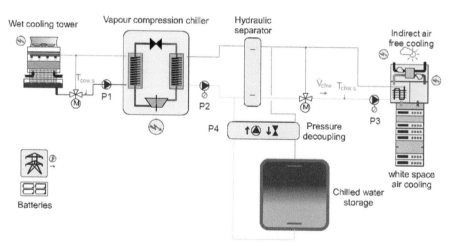

Figure 6.14 Thermal scheme of concept 4.

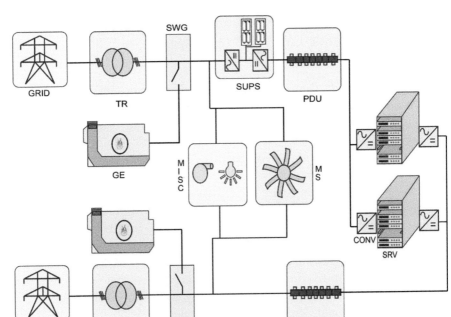

Figure 6.15 Electric scheme of concept 4.

Table 6.6 Subsystems in concept 4

Subsystem	Comments
Lithium-ion battery	Sized according to the required electrical power for IT and cooling distribution as well as the envisaged charging and discharging time
Vapour-compression chiller	Highly efficient machine, e.g., with screw compressor or turbo compressor; backup chiller used for storage charging
Wet cooling tower	Sized according to the chiller capacity
Chilled water storage tank	Non-pressurised tank; contains advanced charging and discharging system for good thermal stratification; sized according to required cooling power and aspired discharging time
Indirect air free cooling	Could include adiabatic cooling; run as often as possible
Hydraulic separator	Separates hydraulic circuits of chillers, storage and consumers (cooling coils)
Pressure decoupling	Decouples storage from network pressure; contains pump for feeding water from the storage into the cooling system and valves for reducing the pressure of water fed into the storage

Operation and Control

Figures 6.16 and 6.17 depict the cooling control strategy and the control strategy of the electrical energy storage of concept 4. Here, the cooling control strategy varies depending on the operating parameters such as the ambient air temperature, electricity price, share of renewable power and the chilled water storage system conditions. The operations of the batteries vary depending on the operating parameters such as cost of the electricity, share of the renewable power and the state of the charge of the batteries. The operating mode of the concept is selected based on the availability of the one of the following order of the operating parameters; first, high share of renewable power in the grid and second cheap electricity. In some scenario, there is a possibility of the availability of more than one operating parameter.

In case that the outside air temperature is lower than the supplied air temperature to the IT room, indirect air free cooling is used. In addition, the chiller can be used to produce chilled water when the share of renewable power is high in the grid or when the electricity cost is low and cold can be stored in the CHWST. In other case that the outside air temperature is higher

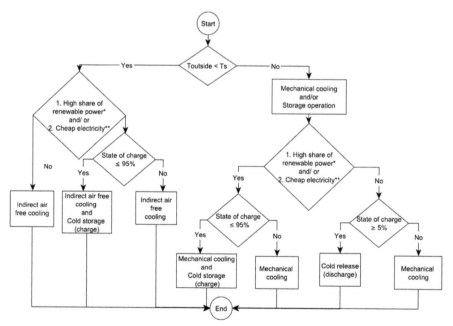

Figure 6.16 Flow chart for the cooling control strategy storage of concept 4.

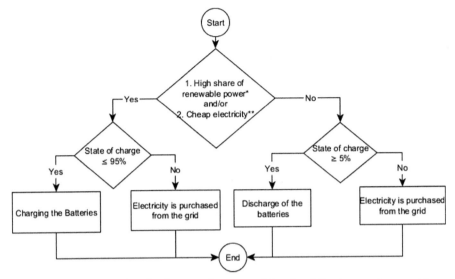

Figure 6.17 Flow chart for the control strategy of the electricity storage of concept 4.

than the supplied air temperature, indirect air free cooling is not applicable. In this scenario, CHWST can partially fulfil the cooling requirements depending on its condition but mainly the vapour-compression chiller produces the cooling energy. Therefore, the vapour-compression chiller is operated interacting with the chilled water storage, based on the cooling energy demand and boundary conditions such as the current electricity cost or share of renewable power, for instance. The set point values for the supply temperatures of the chilled water ($T_{chw,s}$) and the cooling water ($T_{cow,s}$) used in this concept are 10°C and 27°C, respectively. The pump used for the cooling water circuit (i.e. P1) is operated with constant speed, whereas the other pumps (i.e. P2, P3 and P4) are operated with variable speed. The flow rate of fluid through the chiller water pump P3 ($\dot{V}_{chw}$) is controlled in order to maintain a specific supply air temperature of 20°C into the white space.

Limits of Application

Plenty of makeup water is required for the application of wet cooling towers. In addition, local regulation must be considered in order to implement these cooling towers. For the chilled water storage and the electrical storage, sufficient space is required.

Backup and Redundancy

As a redundancy, $N + 1$ compression chillers and wet cooling towers are used to attain redundancy level III. All components are connected by two independent paths.

Sankey Analysis

Here, the location is Frankfurt and the parameters for the simulation from scenario 4 (Table 6.2) are used. It is visible in Figure 6.18 that both the thermal and the electrical storage systems fulfil only the fraction of the total energy required by the Data Centre. About 3% of the imported electricity is stored temporarily in the battery. The share of the indirect air free cooling in the cooling energy demand of the Data Centre is 77%, and the remaining cooling load is supplied by the chiller. About 61%[5] of the energy of the chilled water produced by the chiller is being temporarily stored in the storage, and the rest is used directly to dissipate the waste heat of the AHU unit.

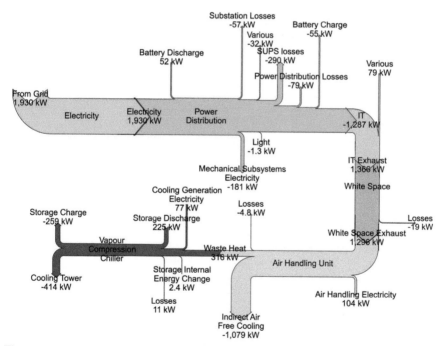

Figure 6.18 Sankey chart showing the distribution of average energy flows per year within different subsystems of concept 4 for scenario 4 (Boundary condition: Frankfurt, 2000 kW IT power capacity, RES = 0.26).

---

[5]For this calculation, a *COP* of the vapour-compression chiller of 5.5 is assumed.

6.2.5 Biogas Fuel Cell with Absorption Chiller

General

A biogas-fed fuel cell is applied for generating both power and heat, which is used for driving an absorption chiller during summer (Figures 6.19 and 6.20). In winter, indirect air free cooling avoids the operation of the chillers. Then, the waste heat from the fuel cell can be recovered for space heating or might also be dissipated by a wet cooling tower. Because of the high temperature and pressure of the hot water, shell and tube heat exchanger are used for transferring the heat between the cooling tower and the fuel cell hot water circuit.

The concept can be realised everywhere where biogas is available. It might be applied in small as well as very large Data Centres (50 kW to >10 MW). Table 6.7 shows an overview of the main components of the concept.

Operation and Control

In summer, the fuel cell is operated according to the heat demand of the absorption chiller. Additional electrical energy can be purchased from the grid. In winter, either the fuel cell can be controlled according to the power demand and its waste heat is dissipated by means of cooling towers, or it is controlled according to the demand of a space and domestic hot water heating. As far as the absorption chiller is concerned, a set point value is defined for the chilled water supply temperature.

Table 6.7 Subsystems in concept 5

Subsystem	Comments
Fuel cell	e.g., SOFC; sized according to the heat demand
Absorption chiller	Double effect, heat supply e.g., at 170/140°C
Wet cooling tower	Sized according to the chiller capacity and for dissipating the waste heat from the fuel cell when there is no heat demand
Hot buffer storage	Allowing for optimal operation both of the heat generator (fuel cell) and the heat consumer (e.g., domestic hot water) by smoothing temperature fluctuations; works as hydraulic separator between generator and consumer circuit
Cold buffer storage	See hot buffer storage
Space heating system	Mixing loop for decreasing the supply temperature in the case of high temperature fuel cell
Indirect air free cooling	Could include adiabatic cooling; run as often as possible
Heat exchanger	Shell and tube heat exchanger; sized according to temperatures of the hot water and mean temperature difference

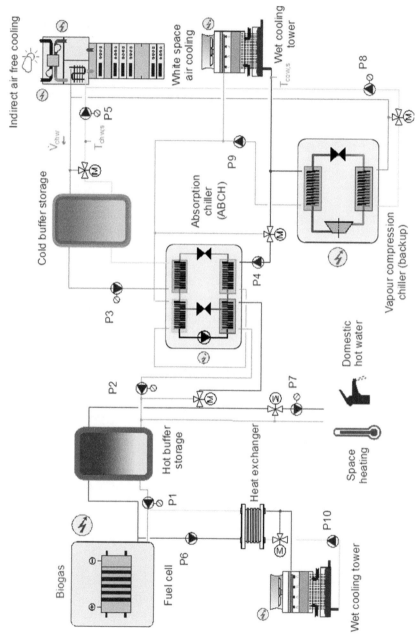

Figure 6.19 Thermal scheme of concept 5.

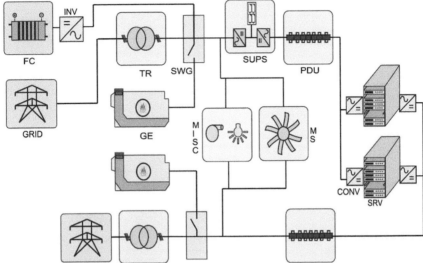

Figure 6.20   Electric scheme of concept 5.

The hot water from the fuel cell is set to 150°C. In addition, set point values are defined for the chilled water supply temperature as well as for the cooling water supply temperature. The set point values for the supply temperatures of the chilled water ($T_{chw,s}$) and the cooling water ($T_{cow,s}$) used in this concept are 10°C and 27°C, respectively. The pumps used for the cooling water circuit (i.e. P4, P6, P9 and P10) are operated with the constant speed, whereas the other pumps are operated with the variable speed pumps. The flow rate of fluid through the chiller water pump P5 ($\dot{V}_{chw}$) is controlled in order to maintain a specific supply air temperature of 20°C into the white space.

The following equations have been used for calculating the amount of the heat and the electricity produced and the fuel consumption of the CHP plant in the simulation of the concept:

$$\dot{Q}_{H,FC} = \dot{Q}_{H,max,FC} \cdot \phi_{max}$$

$$P_{el,FC} = P_{el,max,FC} \cdot \phi_{max}$$

$$\phi_{max} = \frac{\dot{Q}_{H,Abch} + \dot{Q}_{H,DHW}}{\dot{Q}_{H,max,FC}}$$

$$\dot{Q}_{Fuel} = \frac{\dot{Q}_{H,FC} + P_{el,FC}}{\eta_{FC}}$$

where $\dot{Q}_{H,FC}$ represents the heat generated by the fuel cell, $\dot{Q}_{H,max,FC}$ represents the maximum heat generated by the fuel cell, $P_{el,FC}$ represents the electricity produced by the fuel cell, $P_{el,max,FC}$ represents the maximum electricity produced by the fuel cell, $\dot{Q}_{H,Abch}$ represents the heat required by the absorption chiller to produce the cooling energy, $\dot{Q}_{H,DHW}$ represents the heat supplied to the space heating, ϕ_{max} is the maximum load factor, $\dot{Q}_{Fuel}$ is fuel input in CHP system, and η_{FC} is the total efficiency of the fuel cell.

Similarly, the following strategies have been followed for the dissipation of the generated heat by the fuel cell.

$\dot{Q}_a = \dot{Q}_{H,FC} - \dot{Q}_{H,Abch}$, where $\dot{Q}_a$ is the excess heat after supplying for the chiller
If $\dot{Q}_a > 0$,
$\dot{Q}_b = \dot{Q}_a - \dot{Q}_{H.DHC}$, where $\dot{Q}_b$ is the excess heat after supplying for the space heating
If $\dot{Q}_b > 0$,
$\dot{Q}_{H,WCT} = \dot{Q}_b$, where $\dot{Q}_{H.WCT}$ is the excess heat dissipated by the cooling tower.

Limits of Application

The applicability of this concept is limited by the availability of biogas and makeup water for the wet cooling tower as well as local legislation for such cooling towers. Because of the high electrical efficiency of fuel cells, a heat demand for heat reuse is not essential.

Backup and Redundancy

Redundancy level III can be reached by installing N vapour-compression chillers as backup (in the case of any maintenance or failure of the steam plant or absorption chillers). $N + 1$ wet cooling towers are installed, which are connected to both the absorption chillers and vapour-compression chillers. All components are connected by two independent paths. The required electrical power can be purchased from the national grid when the power from the fuel cell falls short.

Sankey Analysis

The location for the simulation is Frankfurt. The parameters applied for the simulation are used from scenario 4 (Table 6.2). Figure 6.21 shows the

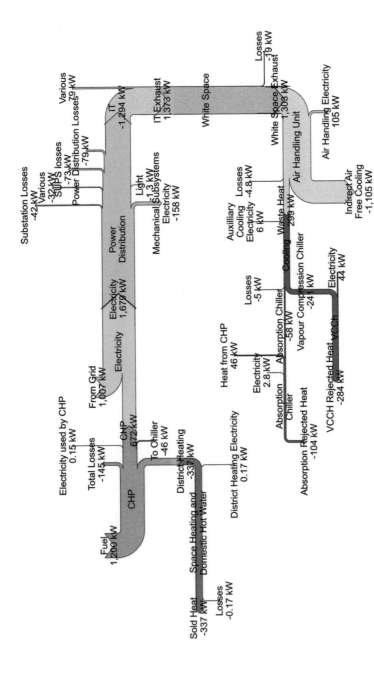

Figure 6.21 Sankey chart showing the distribution of average energy flows per year within different subsystems of concept 5 for scenario 4 (Boundary condition: Frankfurt, 2000 kW IT power capacity, RES = 0.26).

distribution of the average energy flows per year within the different subsystems of concept 5. The excess electricity produced by the CHP system is fed in to the national grid. The CHP plant generates 56% electricity and 32% heat. Only 12% of the useful heat produced by the CHP system is supplied to the absorption chiller to produce the cooling energy and the rest, i.e., 88% of the useful heat, is used for the space heating and domestic hot water. The reason for this is the location, which enables the free cooling to provide a majority of the heat removal, which in turn lowers the cold demand and thereby the heat use by the absorption chiller from the CHP. It is visible that the backup vapour-compression chiller in the system is still needed occasionally to cover certain partial load situations. The free cooling energy covers about 79% of the total cooling energy demand of the Data Centre.

6.2.6 Reciprocating Engine CHP with Absorption Chiller

General

The concept 6 shown schematically in Figures 6.22, 6.23 and 6.24 is based on biogas-fed tri-generation by means of a reciprocating engine CHP plant. The heat from this plant is used for driving a single-effect absorption chiller during summer and supplying space heating for offices or buildings close to the Data Centre during winter. Additionally, indirect air free cooling is implemented for efficient cooling supply to the Data Centre especially during winter. Then, the heat from the CHP plant should be used for space heating and producing domestic hot water if required (generally, hot water demand is quite low in offices).

The concept is not subject to any geographical restrictions and can be realised everywhere where biogas is available. It might be applied in very small as well as large Data Centres (50 kW to 10 MW).

Table 6.8 gives an overview on the main components (subsystems) shown in Figure 6.22.

Operation and Control

Generally, the CHP plant is operated according to the heat demand for driving the chiller or supplying heat for space heating. If the generated power does not match the current power demand of the Data Centre, additional power has to be purchased from the national grid or excess power can be sold to the grid.

If the implementation of the cooling system has multiple cooling towers and chillers, then for the efficient part load operation, sequenced operation

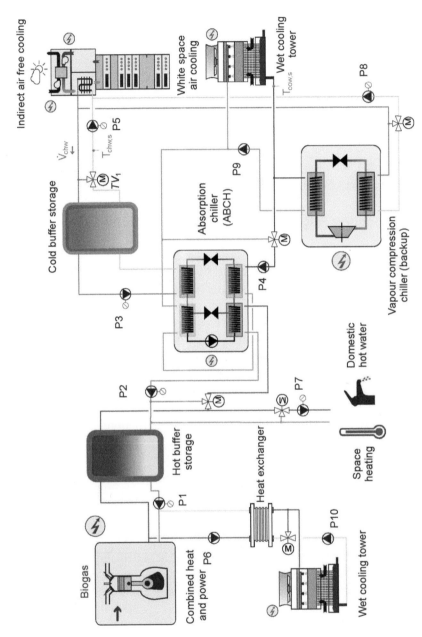

Figure 6.22 Thermal scheme of concept 6.

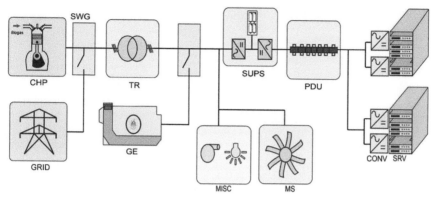

Figure 6.23 Electric scheme of concept 6 with off-site generation.

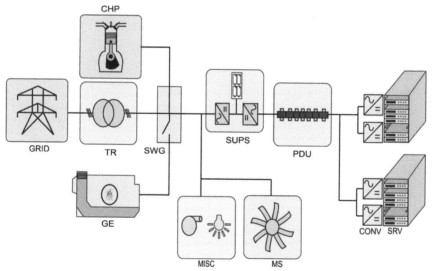

Figure 6.24 Electric scheme of concept 6 with on-site generation.

strategies should be used. Hot water of around 90°C is fed into the chiller. In addition, set point values are defined for the chilled water supply temperature as well as for the cooling water supply temperature. The set point values for the supply temperatures of the chilled water ($T_{chw,s}$) and the cooling water ($T_{cow,s}$) used in this concept are 10°C and 27°C, respectively. The pumps used for the cooling water circuit (P4, P6, P9 and P10) are operated with constant speed, whereas the other pumps are operated with variable speed. The flow

Table 6.8   Subsystems in concept 6

Subsystem	Comments
Reciprocating engine CHP plant	Sized according to the heat demand (supplying absorption chiller in summer and space heating in winter)
Absorption chiller	Single-effect, heat supply, e.g., at 85/70°C
Wet cooling tower	Sized according to the chiller capacity and for dissipating the waste heat from the CHP when there is no heat demand
Heat exchanger	Shell and tube heat exchanger; sized according to temperatures of the hot water and mean temperature difference
Hot buffer storage	Allowing for optimal operation both of the heat generator (CHP plant) and the heat consumer (e.g., chiller) by smoothing temperature fluctuations; works as hydraulic separator between generator and consumer circuit
Cold buffer storage	See hot buffer storage
Space heating system	Design supply temperature adapted to the temperature level provided by CHP, e.g., 70°C
Indirect air free cooling	Could include adiabatic cooling; run as often as possible

rate of fluid through the chiller water pump P5 ($\dot{V}_{chw}$) is controlled in order to maintain a specific supply air temperature of 20°C into the white space.

The following equations have been used to calculate the amount of heat and electricity produced and the fuel consumption of the CHP plant in the simulation of the concept:

$$\dot{Q}_{H,eff} = \dot{Q}_{H,max,CHP} \cdot \phi_{max}$$

$$P_{el,eff} = P_{el,max,CHP} \cdot \phi_{max}$$

$$\phi_{max} = \frac{\dot{Q}_{H,Abch} + \dot{Q}_{H,DHC}}{\dot{Q}_{H,max,CHP}}$$

$$\dot{Q}_{Fuel} = \frac{\dot{Q}_{H,eff} + P_{el,eff}}{\eta_{CHP}}$$

where $\dot{Q}_{H,eff}$ represents the heat generated by the CHP, $\dot{Q}_{H,max,CHP}$ represents the maximum heat generated by the CHP, $P_{el,eff}$ represents the electricity produced by the CHP, $P_{el,max,CHP}$ represents the maximum electricity produced by the CHP, $\dot{Q}_{H,Abch}$ represents the heat required by the absorption chiller to produce the cooling energy, $\dot{Q}_{H,DHC}$ represents the heat supplied to the space heating, ϕ_{max} is the maximum load factor, $\dot{Q}_{Fuel}$ is the fuel input in the CHP system, and η_{CHP} is the total efficiency of the CHP plant.

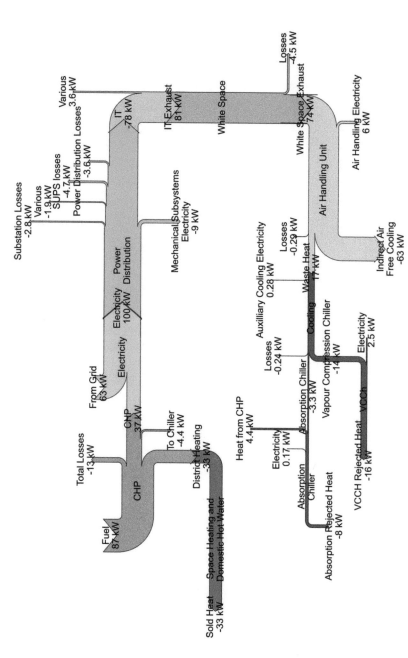

Figure 6.25 Sankey chart showing the distribution of average energy flows per year within different subsystems of concept 6 for scenario 2 (Boundary condition: Frankfurt, 120 kW IT power capacity, RES = 0.26).

Limits of Application

The concept is suitable for all kinds of Data Centres. However, the CHP plant requires a certain amount of annual operating hours in order to make economic sense. Thus, there must be an appropriate heat demand available close to the Data Centre, which absorbs the heat especially during winter, when the Data Centre is cooled by means of indirect air free cooling.

Backup and Redundancy

A backup gas boiler can generate heat for the chiller when the CHP plant fails. Redundancy level II can be reached by installing $N + 1$ chillers and cooling tower units.

Sankey Analysis

For the simulation, a 120 kW IT power capacity Data Centre located in Frankfurt and the parameters from the scenario 2 (Table 6.2) are applied. Figure 6.25 shows the distribution of the average energy flows per year within the different subsystems of concept 6. The excess electricity produced by the CHP system is fed into the national grid. The chart shows that the indirect free cooling covers around 79% of the total cooling energy demand of the Data Centre and the rest is covered by the chillers. The CHP plant operates with an electrical efficiency of 43% and a thermal efficiency of 43%. Only 12% of the useful heat produced by the CHP system is used to the absorption chiller to produce cold, and the rest is fed in to the space heating. The operation of the absorption chiller and vapour-compression chiller and the issues are the same as explained for concept 5.

References

[1] Agència de l'Energia de Barcelona. *Dades facilitades sobre el recurs eòlic disponible a la ciutat de Barcelona.*

[2] Intelligent Energy Europe programme. *Catalogue of European Urban Wind Turbine Manufacturers.*

[3] (2015). Retrieved from http://www.deltapowersolutions.com/en-in/20kva-and-higher-modular-ups-modulon-family.php

[4] 451 Research Data center technologies. (May 2013). *Disruptive Technologies in the Data Center-Ten Technologies Driving a Wave of Change.*

[5] Acharya, P., Enjeti, P., and Pitel, I. (2004). An advanced fuel cell simulator. In *Applied Power Electronics Conference and Exposition, 2004. APEC '04. Nineteenth Annual IEEE*, 3 (4), 1554–1558.

[6] ASUE. (2011). *BHKW-Kenndaten* 2011. Berlin: energieDRUCK.

[7] Bahnfleth, W. P., and Peyer, E. (2004). *Variable primary flow chilled water systems: Potential benefits and application issues*. State College: The Pennsylvania State University.

[8] Cace, J., Horst, E.t., and Syngellakis, K. (2007). *Guidelines for Small Wind Turbines in the Built Environment*. Intelligent Energy-Europe.

[9] Carnegie, R., Gotham, D., Nderitu, D., and Preckel, P. V. (2013). *Utility scale energy storage systems: benefits, applications and technologies*.

[10] Cemep and Assoautomazione. ANIE European Guide.

[11] CLIMAVENETA. (n.d.). TECS2/XL-CA 0351.

[12] Diaz-Gonzalez, F., Sumper, A., Gomis Bellmunt, O., and Villafafila Robles, R. (2012). A review of energy storage technologies for wind power applications. *Renewable and Sustainable Energy Reviews*, 16, 2154–2171.

[13] DIN V 18599-7. (2011). Energy efficiency of buildings – Calculation of the net, final and primary energy demand for heating, cooling, ventilation, domestic hot water and lighting. Berlin: Deutsches Institut für Normung e.V.

[14] Dincer, I., and Kanoglu, M. (2011). *Refrigeration Systems and Applications*. Chichester: John Wiley and Sons.

[15] EASE/EERA. (2013). *Joint EASE/EERA recommendations for an European Energy Storage Technology Development Roadmap towards 2030*.

[16] Ecotermia. (n.d.). *Electricity especific factors for grid electricity*. Retrieved 2015, from http://ecometrica.com/assets/Electricity-specific-emission-factors-for-grid-electricity.pdf

[17] Eurostat. (n.d.). *Eurostat Statistics Explained*. Retrieved 2015, from http://ec.europa.eu/eurostat/statistics-explained/index.php/Energy_price_statistics

[18] Felman, D., and Barbose, G. (November 2012). *Photovoltaic (PV) Pricing Trends: Historical, Recent, and Near-Term Projections*. Technical Report DOE/GO-102012-3839.

[19] FuelCell Energy. (2013). (FuelCell Energy) Retrieved from http://www.fuelcellenergy.com/why-fuelcell-energy/benefits/

[20] Gebhardt, M., Kohl, H., and Steinrötter, T. (2002). *Preisatlas – Kostenfunktionen für Komponenten der rationellen Energienutzung*. Duisburg: Institut für Energie- und Umwelttechnik e.V. (IUTA).

[21] Henning, H.-M., Urbaneck, T., et al. (2009). *Kühlen und Klimatisieren mit Wärme*. Stuttgart: Fraunhofer IRB Verlag.

[22] Hirschenhofer, J. (1992). How the fuel cell produces power. *IEEE Aerospace & Electronic Systems Magazine,* 11 (2), 24–25.

[23] Hirschenhofer, J., Staffer, D., Englemann, R., and Klett, M. (1998). *Fuel Cell Handbook*. USA: Parsons Corporation of US. Department of Energy, Office of Fossil Energy.

[24] HYDROGENICS. (2013). *hydrogenics*. Retrieved from http://www.hydrogenics.com/docs/default-source/default-document-library/hypm-hd-180.pdf?sfvrsn=0

[25] IEA Energy Technology Essentials. (2007). *Fuel Cells*. France.

[26] IEA-ETSAP and IRENA. (2013). *Heat pumps – technology brief*.

[27] IKET (Ed.). (2005). *Pohlmann-Taschenbuch der Kältetechnik*. Heidelberg: Müller.

[28] JRC Euopean Commission. (n.d.). *http://re.jrc.ec.europa.eu/pvgis/*. Retrieved 2014.

[29] Karady, G. (MAy, 2002). *Investigation of Fuel Cell System Performance and Operation*. Arizona State University, USA: PSERC Publication.

[30] McCarthy, N. (n.d.). *White Paper # 75*. (APC Scheider Electric) Retrieved from http://www.apcmedia.com/salestools/SADE-5TPL8X/SADE-5TPL8X_R3_IT.pdf

[31] Meyer, J. P. (2011). Heat Pumps. (Thermopedia) Retrieved from thermopedia.com/content/837

[32] Munters. Indirect and Direct Evaporative Cooling. (n.d.). Retrieved from http://www.munters.us/en/us/Products–Services/Dehumidification/Indirect-Heating–Cooling/Indirect-Cooling/

[33] National Renewable Energy Laboratory (NREL). *Cost and Performance Data for Power Generation Technologies*.

[34] National Renewable Energy Lboratory. Gas-fired distributed energy resource technology characterizations. (2003, November). Retrieved from http://www.nrel.gov/analysis/pdfs/2003/2003_gas-fired_der.pdf

[35] Nehrir, H., and Wang, C. (2009). Modelling and Control of Fuel Cells: Distributed Generations Applications. *IEEE Press Series on Power Engineering*.

[36] Philips, R. (2009). Using Direct Evaporative + Chilled Water Cooling. *ASHRAE Journal*.

[37] Piller Corporation. (2014). Retrieved from http://www.piller.com/

[38] Rasmussen, N. (n.d.). *White Paper 1 Rev 7.* (APC Scheider Electric) Retrieved from http://www.apcmedia.com/salestools/SADE-5TNM3Y/SADE-5TNM3Y_R7_EN.pdf

[39] RenewIT. (2015). *Catalogue of advanced technical concepts for Net Zero Energy Data Centres.*

[40] RenewIT. (2014). *Data Centres: Markets, archetypes and Case Studies.*

[41] RenewIT. (2014). *Energy requirements for IT equipment.*

[42] RenewIT. (2014). *Report of different options for renewable energy supply in Data Centres in Europe.*

[43] Riekstin, A. C., James, S., Kansal, A., Liu, J., and Peterson, E. (2013, November). No more Electrical Infrastructure: Towards Fuel Cell Powered Data Centers. *Hot Power 13.*

[44] Scofield, M., and Dunnavant, K. (1999). Evaporative Cooling and TES after deregulation. *ASHRAE J, 41* (12), 31–36.

[45] Solair Project. (2009). *Requirements on the design and configuration of small and medium-size solar air-conditioning applications.* Freiburg: Fraunhofer ISE.

[46] St. Lawrence University. (n.d.). Retrieved from http://it.stlawu.edu/~jahncke/clj/cls/317/EnergyEfficiency.pdf

[47] Stiller, C. (May 2006). Doctor Thesis: Design, Operational & Control Modelling of SOFC/GT Hybrid Systems. Trondheim, Norway: Norwegian University of Science and Technologies.

[48] TECHNOwind. (n.d.). *http://www.technowind.eu/.* Retrieved 2014.

[49] Thermea. (2010). Hochtemperaturwärmepumpen und Kältemaschinen mit dem natürlichen Arbeitsstoff CO2. dresden: thermea. energiesysteme GmbH.

[50] TRNSYS. (2012). *TRNSYS 17 Mathematical Reference.* Madison: TRNSYS.

[51] U.S. Environmental Protection Agency. (2008). *Catalog of CHP Technologies.*

[52] Urbaneck, T. (2014). Internal information. Chemnitz: Technische Universität Chemnitz.

[53] Urbaneck, T. (2012). *Kältespeicher.* München: Oldenbourg Wissenschaftsverlag.

[54] Wagner, W. (1993). *Wärmeaustauscher.* Würzburg: Vogel.

7

Applying Advanced Technical Concepts to Selected Scenarios

Verena Rudolf[1], Nirendra Lal Shrestha[1], Eduard Oró[2],
Thorsten Urbaneck[1] and Jaume Salom[2]

[1]Chemnitz University of Technology,
Professorship Technical Thermodynamics, Germany
[2]Catalonia Institute for Energy Research – IREC, Spain

7.1 Overview of Concept Performance

In this section, the energetic and economic performance of the concepts are analysed varying the main Data Centre's characteristics. To do so, the following metrics are used (see Chapter 3 for a detailed description):

- Normalized[1] non-renewable Data Centre primary energy ($PE_{DC,nren}/ Nominal\ IT\ Power$)
- Normalized CAPEX ($CAPEX/Nominal\ IT\ Power$)
- Normalized OPEX ($OPEX/Nominal\ IT\ Power$)
- Normalized total cost of ownership ($TCO/Nominal\ IT\ Power$)
- Power usage effectiveness (PUE)
- Renewable energy ratio (RER)

TRNSYS [1] simulations over an entire year are used to investigate the performance of each concept. This software is mainly used to model and simulate systems that are influenced by several independent factors and that involve non-cyclical storage processes. It offers a broad variety of standard components such as pumps, buildings, wind turbines, weather data processor etc. and has capabilities to import components from other libraries, for example, TESS, Transsolar, etc. The performance of the models with respect to the variation of some of the most important parameters such as the location

---

[1]A normalized metric means that the standard metric is divided by the IT power capacity of the Data Centre in kW.

and size of the Data Centre is investigated using the Monte Carlo[2] sampling of the parameter space. For this, 100 simulations with a duration of one year were performed in order to generate the results. As described, one of the most important parameters is the location, which is one of the driving force behind the results. The location determines many input variables of the simulation that affects the system performance. The most important ones are:

- Environmental conditions, strongly related with free cooling potential;
- Electricity price profile;
- Share of renewables in the electricity grid.

Therefore, in this section, the performance of each concept is investigated considering the different locations. For the analysis of the concepts, 26 different locations within Europe have been considered (Table 7.1). These European cities were selected taking into account the climate zones as proposed in Köppen climate classification [2], such as dry, temperate, continental, polar and tropical climates. During the simulation, the other parameters of each concept are unchanged.

Figure 7.1 to Figure 7.6 show the box plots[3] for the different metrics. Different colours categorize the concepts. Green is used for the concept with renewable energy sources (concept 1), purple shows the concept with thermal end electrical storage (concept 4), red is used for concepts with CHP systems (concepts 5 and 6) and the other concepts are marked in blue (concepts 2 and 3).

Notice that the energy metrics would change drastically in function of the fuel used to produce the cooling in the district heating and cooling plant or the fuel used in the CHP plant. Table 7.2 shows the average normalized ratios for primary energy for European countries.

Figure 7.1 shows the boxplots of the distribution of the non-renewable primary energy consumption per nominal IT power for each advanced solution analysed. The higher value of this metric represents the higher impact on the environment. The median value of the primary energy consumption is different for all the concepts. Concept 4 has the highest median values indicating the highest primary energy consumption, because the concept does not incorporate any renewable energy generation, only storage (both, electrical and thermal). Concepts 5 and 6 have the lowest values, because

---

[2]Monte Carlo method is a broad class of computational algorithm that rely on repeated random sampling to obtain numerical results.

[3]Boxplots are a very compact way to visualize data. The dots represent outliers. The vertical line at the bottom is the bottom quartile of the data, the box itself covers the 2nd and 3rd quartile and the vertical line on top of the box covers the top quartile of the data.

Table 7.1 European cities selected for the concept analysis

City		Country	Climate Zone
Almeria	ALM	Spain	Dry
Madrid	MAD	Spain	Dry
Valencia	VAL	Spain	Dry
Bergen	BGO	Norway	Polar and Alpine
Innsbruck	INN	Austria	Polar and Alpine
Zurich	ZRH	Switzerland	Polar and Alpine
Belgrade	BEG	Serbia	Continental
Kaunas	KUN	Lithuania	Continental
Kiev	KBP	Ukraine	Continental
Stockholm	STK	Sweden	Continental
Warsaw	WAW	Poland	Continental
Amsterdam	AMS	The Netherlands	Temperate
Barcelona	BCN	Spain	Temperate
Edinburgh	EDI	Scotland	Temperate
Frankfurt	FRA	Germany	Temperate
Milano	MIL	Italy	Temperate
London	LON	UK	Temperate
Paris	PAR	France	Temperate
Porto	OPO	Portugal	Temperate
Seville	SVQ	Spain	Temperate
Chemnitz	CHE	Germany	Temperate
Roma	ROM	Italy	Temperate
Verona	VER	Italy	Temperate
Groningen	GRO	Netherlands	Temperate
Rotterdam	ROT	Netherlands	Temperate
Ancona	ANC	Italy	Temperate

Table 7.2 Primary and final energy and CO_2 emissions conversion factors

Energy Source	Non-Renewable Primary Energy Weighting Factor [3]	Total Primary Energy Weighting Factor [3]
PV	0.00	1.00
Wind	0.00	1.00
Biogas	0.50	1.50
Biomass	0.05	1.05
District cooling[4]	0.60	1.70
Exported heat	0.70	2.00

---

[4]District cooling primary Energy and CO_2 emission factors are dependent of the energy used and the district heating and cooling network. In the study, the factors are the ones from Parc de l'Alba [4] district network.

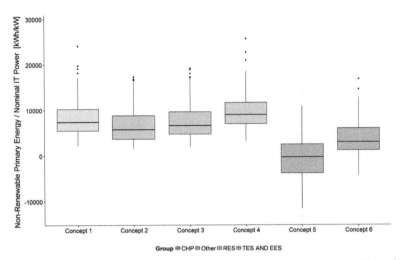

Figure 7.1   Evaluation of the concepts with respect to the non-renewable primary energy/nominal IT power.

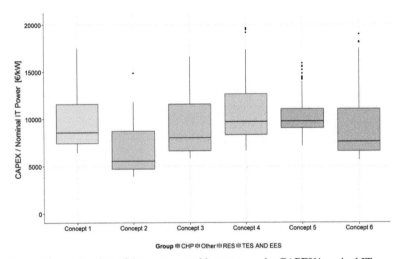

Figure 7.2   Evaluation of the concepts with respect to the CAPEX/nominal IT power.

the energy generated by the CHP plant has a very low non-renewable primary energy factor due to the fuels used and a complete re-use of the excess of heat produced by the CHP and not used by the Data Centre itself. The negative value of the primary energy consumption (concept 5) represents bigger generation of the primary energy than required for the Data Centre and thus, an export of the excess energy. Therefore, since the excess of energy is exported, concept 1,

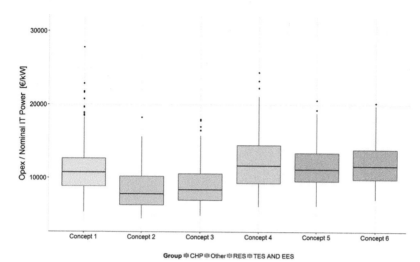

Figure 7.3   Evaluation of the concepts with respect to the OPEX/nominal IT power.

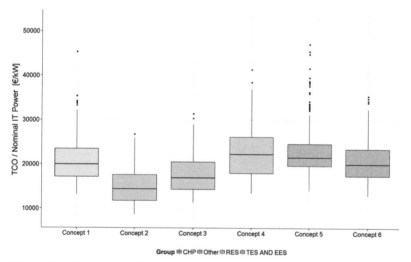

Figure 7.4   Evaluation of the concepts with respect to the TCO/nominal IT power.

even with a high renewable energy generation, has higher primary energy consumption than the concepts with the CHP system.

It can be seen in the Figure 7.2 that the investment cost of the systems (CAPEX) with a higher renewable energy generation (concepts 1 and 5) and with thermal and electrical storage (concept 4) is higher. The lowest investment cost is shown for concept 2, because district heating and cooling is

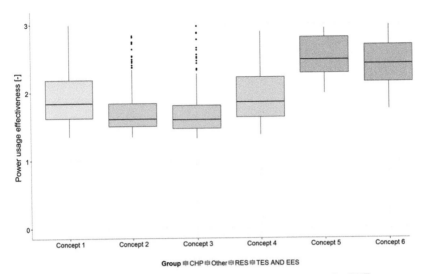

Figure 7.5 Evaluation of the concepts with respect to the PUE.

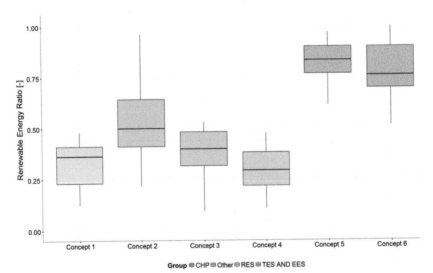

Figure 7.6 Evaluation of the concepts with respect to the RER.

an established technology and the Data Centre owner just pay for the con-
nexion. Analysing the operational cost (Figure 7.3) of the systems (OPEX),
concept 2 and 3 are the ones with lower cost of operation. On one hand,
the cooling provided by the district cooling network is cheaper than the one

generated onsite and on the other hand, the use of free cooling helps to reduce drastically the energy consumption for the cooling system. Concepts with CHP technologies have higher operational costs since they need biogas to run the engines and the fuel cells. With the addition of the operational cost (OPEX) to the capital cost (CAPEX), concept 2 followed by concept 3 are observed to be the better systems in term of the lowest global cost (TCO) per nominal IT power consumption (Figure 7.4).

Figure 7.5 shows the median value of the PUE for the different concepts. Note that for concepts with CHP systems (concepts 5 and 6) the median PUE is more than 2. This is because the rejected heat of the system is not counted in the PUE calculation, as the standardization bodies establish. Therefore, PUE metric for these systems does not capture the essence of such systems. In a Data Centre, excess heat could be from a CHP plant or from the IT white space. The re-used waste heat from IT is assumed to be 0 in the PUE calculation. On the other hand, the other concepts, especially concept 2 and 3 present median PUE values aligned with current trends. In order to make the Data Centre more energy efficient, an optimization process is needed for each specific concept and location.

Figure 7.6 illustrates the RER for the different concepts. Advanced solutions with the CHP system (concepts 5 and 6) have a higher RER, followed by the solution connected to the district cooling (concept 2).

7.2 Concept Comparison for Selected Scenarios

7.2.1 Description of Scenarios Analysed

The objective of this section is to study the energy and economic feasibility of different advanced concepts in three different locations (Barcelona, Stockholm and Frankfurt) under the same boundary conditions. It is considered a Data Centre with an IT power capacity of 1000 kW. The other main parameters such as the rack density, the occupancy ratio, the safety margin factor and the load profile are fixed as shown in Table 7.3. The safety margin factor is used to limit the maximum IT power capacity that the servers can run in the installation. As an example, a safety margin of 0.8 means that the maximum IT power consumption would be 80% of the total IT power capacity (1000 kW). The occupancy ratio means the ratio of installed IT, lack of occupancy is a lack of IT equipment. Therefore, for a Data Centre (power capacity of 1000 kW) with a safety margin of 0.8 and an occupancy ratio of 0.5, the maximum IT power consumption of the servers is 400 kW. The white space area basically

Table 7.3   Specific assumptions for the investigated concepts for a 1000 kW IT power Data Centre in three locations Barcelona (BCN), Stockholm (STO) and Frankfurt (FRA)

Parameter	Unit	BCN	STO	FRA
Location	[–]	Barcelona	Stockholm	Frankfurt
IT power capacity	[kW]	1000	1000	1000
Rack density	[kW/rack]	4	4	4
Occupancy ratio	[–]	1	1	1
Safety margin factor	[–]	0.8	0.8	0.8
White space area	[m^2]	750	750	750
IT Load profile	[–]	Mixed	Mixed	Mixed
Average electricity price	[€/kW·h$_{el}$]	0.0988	0.0630	0.0720
Biogas price	[€/kw·h$_{biogas}$]	0.08	0.08	0.08
District cooling price	[€/kW·h$_{DCool}$]	0.035	0.035	0.035
Exported heat price	[€/kW·h$_{heat}$]	0.025	0.025	0.025
Average ratio renewables in the grid	[–]	0.36	0.62	0.26
$w_{del,total,el}$ (average)	[–]	2.29	1.86	2.32
$w_{del,nren,el}$ (average)	[kW·h$_{PE}$/kW·h$_{el}$]	1.83	1.30	2.15
$w_{del,nren.biogas}$	[kW·h$_{PE}$/kW·h$_{biogas}$]	0.5	0.5	0.5
$w_{del,nren,DCool}$	[kW·h$_{PE}$/kW·h$_{DCool}$]	0.6	0.6	0.6
$w_{exp,nren,heat}$	[kW·h$_{PE}$/kW·h$_{heat}$]	0.7	0.7	0.7

depends on the nominal IT power capacity (kW) and the rack density (kW per rack). The white space area was estimated using well-known industry average ratios for occupied floor occupied by a rack. The tables in section 3 show some of the basic parameters used to define the sizing of the main elements in the different energy concepts. In this analysis, the IT load profile used is a combination of the most standard IT workload profiles: Web, HPC and Data. In particular, the workload profile used is composed of 35% HPC, 30% Data and 35% Web based on Carbó et al. [5].

The simulation models developed allow the introduction of a set of energy efficiency measures individually or combined. The energy efficiency strategies are technical solutions that can be applied in almost all the Data Centres and combined with any system to supply cooling and power with or without renewable energy sources (RES). First, the strategies that allow reducing the load demand as much as possible have been integrated and analysed. Second, the use of RES has been studied. Energy efficiency measures can be grouped in the following categories:

- Advanced measures for building design. The building design may affect the cooling demand of the Data Centre.
- Advanced measures for electrical supply. Some well-known strategies are modular UPS and bypassed UPS which achieve a reduction of electrical losses in the power distribution lines. In the results presented here, modular UPS have been applied when energy efficiency measures are mentioned.
- Advanced measures for cooling supply. These measures include the use of free cooling, hot/cold aisle containment for a better air management, variable air flow and the increase of allowable IT working temperatures. Using highly energy efficient components, in particular vapour compression chillers and CRAH units, can also lead to a significant reduction of the total energy demand.
- Advanced measures for IT management. Consolidation aims to concentrate IT workloads in a minimum number of servers to maintain the inactive servers in idle state. Then, those servers in idle state can be turned off. Finally, IT scheduling aims to move IT jobs according to the availability of RES when it is possible.

Figure 7.7 and Figure 7.8 present the results of the different advanced concepts proposed at the three different locations. In the same graph the results for each of the concepts applying the complete set of energy efficiency measures are presented and compared with the results of the reference case where none of these energy efficiency measures are implemented. In Figure 7.7 results in the graphs are grouped for each technical concept presenting normalized TCO and RER versus normalized primary energy consumption, respectively left and right. In Figure 7.8 the results are grouped for each location showing in the same graph the results for each concept.

Talking about costs, one can observe that for the same concepts and the same sizes there are significant differences between operating a Data Centre in Barcelona, in Frankfurt or in Stockholm under the hypothesis used in this study. TCO costs are mainly driven by the sizes of the main elements and labour costs of building a Data Centre to determine the CAPEX (which is not influenced by the location since average European prices have been used) and for the energy prices that influence the OPEX (which differences come from the differences in the electricity prices). The average difference of TCO between having a Data Centre in Frankfurt compared to Barcelona is 7% and 24% if Stockholm is compared with Barcelona, Barcelona the location with the highest TCO. This is mainly due to the fact that the electricity price is higher in Barcelona than in Stockholm, while for Frankfurt an intermediate value is reached. As expected, these differences in TCO are reduced in the concepts where less electricity is needed to run the facility, like concept 5 and concept 6 where biogas is used to provide power/cold. In that case, it is still expensive to build and operate a Data Centre in Barcelona but only 3% more expensive compared to Frankfurt and 6% compared to Stockholm. The results present significant differences in the absolute values of PE_{nren} between the three locations. As PE_{nren} considers the amount of non-renewable primary

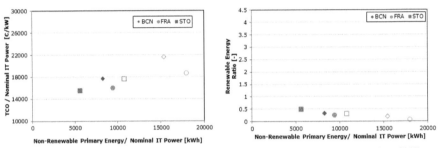

(a) Concept 1. Photovoltaic System and Wind Turbines with Vapour-Compression Chiller and Lead-Acid Batteries.

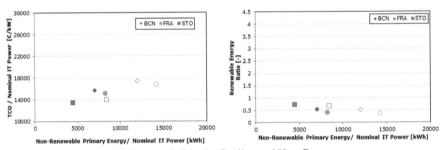

(b) Concept 2. District Cooling and Heat Reuse.

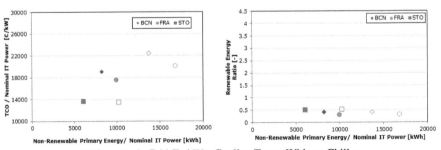

(c) Concept 3. Grid-Fed Wet Cooling Tower Without Chiller.

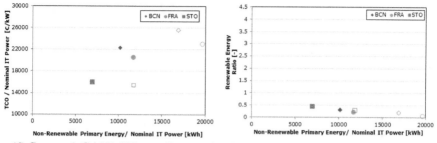

(d) Concept 4. Grid-Fed Vapour-Compression Chiller with Electrical Energy and Chilled Water Storages.

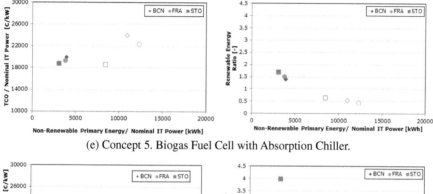

(e) Concept 5. Biogas Fuel Cell with Absorption Chiller.

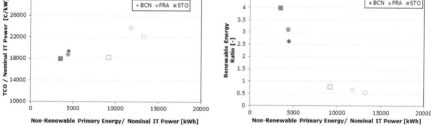

(f) Concept 6. Reciprocating Engine CHP with Absorption Chiller.

Figure 7.7 Normalized TCO vs. normalized PE$_{nren}$ (left) and RER (RER) vs. normalized PE$_{nren}$ (right) for each advanced technical concepts in three different locations (Barcelona, Stockholm and Frankfurt). Results of not applying energy efficiency measures (unfilled shapes) and applying energy efficiency measures (filled shapes) are shown for all the concepts.

energy in the electricity network through the appropriate weighting factors, there is a difference between the locations as well as the influence of other climatic conditions: for example there are more hours of free cooling available in Stockholm than in Barcelona. Table 7.4 shows the results for concept 1

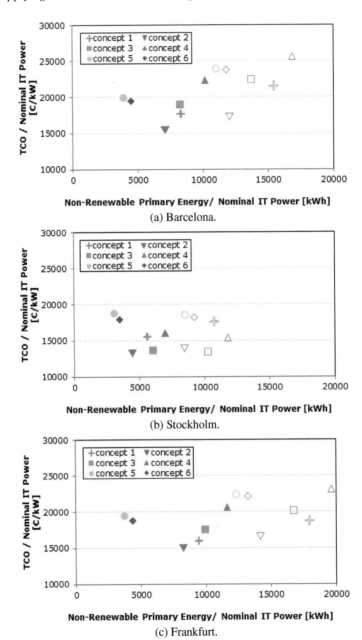

(a) Barcelona.

(b) Stockholm.

(c) Frankfurt.

Figure 7.8 Normalized TCO vs. normalized PE_{nren} for the location of (a) Barcelona, (b) Stockholm and (c) Frankfurt. Results of not applying energy efficiency measures (unfilled shapes) and applying energy efficiency measures (filled shapes) are shown for all the concepts.

Table 7.4 Normalized PE$_{nren}$ consumption for concepts 1 and 5 for the locations of Barcelona, Frankfurt and Stockholm

	BCN	FRA	STO
Concept 1-Convectional DC with VCCH			
Without energy efficiency measures	15 403	17 924	10 776
With energy efficiency measures	8 286	9 419	5 608
Concept 5-Biogas Fuel Cells + Absorption chiller			
Without energy efficiency measures	10 962	12 287	8 478
With energy efficiency measures	3 934	3 786	3 099

and concept 5, where differences between locations in normalized PE$_{nren}$ are reduced because concept 5 is mainly based on biogas which has the same conversion factor for the two locations.

For all the cases analysed, there is a significant benefit of applying as much energy efficiency measures as possible which will produce savings in the total costs and in the primary energy consumption. IT management strategies are the most beneficial ones, together with some measures which allow increasing the total number of free-cooling hours or improving the efficiency in the electrical distribution. For the case studies presented here, the impact of applying energy efficiency measures is a reduction in PE$_{nren}$ up to 48% in Barcelona and 50% in Frankfurt and Stockholm as well as a reduction of TCO up to 15% in Barcelona and 13% in Frankfurt and Stockholm.

According to the results, operating a conventional Data Centre without renewables (concept 1) can cost between 17688 and 21613 €/kW$_{IT}$ and consuming about 15403 kW·h$_{PE,nren}$/kW$_{IT}$·year in Barcelona, 17924 kW·h$_{PE,nren}$/kW$_{IT}$·year in Frankfurt and 10776 kW·h$_{PE,nren}$/kW$_{IT}$·year in Stockholm. Although, it was commented that a significant reduction of TCO and PE$_{nren}$ is possible applying different energy efficiency strategies, a reduction of primary energy resources is possible with different concepts to run a Data Centre. Among the ones detailed described in Chapter Six, concept 5 based on biogas fuel-cells gives the best results in terms of PE$_{nren}$ reduction although it is an expensive concept compared to a conventional Data Centre (concept 1). Having a CHP with a biogas engine (concept 6) gives also promising PE$_{nren}$ savings but is less expensive than concept 5 although still having a higher TCO than a conventional Data Centre. Both concepts 5 and 6 rely also on the availability of biogas as local and/or imported resource and the excess of heat produced by the fuel cells or the CHP system is 100% reused. The most cost effective concept, i.e. the one that combines more

PE$_{nren}$ and TCO savings, is concept 2 which connects the Data Centre to a district cooling system and heat from the Data Centre can be used for heating purposes. The reduction of TCO can reach up to 21% and the PE savings up to 22% when no energy efficiency measures are applied. Concepts 2, 5 and 6 rely on a 100% reuse of the heat produced by the facility. The implementation of concept 3 based on wet cooling towers shows moderate PE savings compared to conventional concept and even no benefit in location where free-cooling strategies can be applied. Concept 4 is the one that shows higher costs and higher primary energy consumption in all of the locations. Higher costs come from the additional storage systems which do not produce enough savings in energy costs to compensate the additional investment. Also, an increase of primary energy consumption is shown. These results related to concept 4 indicate that it is an advanced technical concept which may have some benefits in specific contexts where high fluctuation of energy prices (or share of renewables in the grid) occurs along the day.

7.3 Detailed Analysis by Advanced Technical Concepts

7.3.1 Introduction

In this section, a detailed analysis studying the implementation of energy efficiency measures and varying some main parameters such as the IT power and the size of the elements of the system is presented. The reference system (vapour compression system with CRAH units) is a Data Centre with an IT power capacity of 1000 kW located in two different locations: Barcelona and Stockholm. The main parameters of the installation are shown in Table 7.3. To analyse the influence of the 14 energy efficiency measures accumulative tests (sequential and accumulative implementation of each strategy) have been followed for all the scenarios. The accumulative effect helps to assess the relative influence of the measures and to quantify the optimization. Table 7.5 shows the values used for all the parameters affecting each of the energy efficiency strategy implementation for the accumulative tests.

7.3.2 Concept 1. Photovoltaic System and Wind Turbines with Vapour-Compression Chiller

7.3.2.1 Influence of energy efficiency measures

The results of the accumulative tests are presented in two ways: results for each metric analysed (Figure 7.9) and the sizes of the main elements (Table 7.6).

Table 7.5 Energy efficiency strategies and their values for the accumulation test

Variable Parameter	Unit	01. Reference Case	02. Building Design	03. Bypass UPS	04. Modular UPS	05. Enhanced UPS	06. Thermal Containment	07. Thermal Variable Flow	08. Thermal Indirect Cooling	09. Thermal SupplyAir Temperature	10. Thermal Delta T	11. Thermal High Efficiency	12. IT Consolidation	13. TurnOffIdleServer	14. Green Algorithm
WSP thermal transmissivity	[kJ/ hr·m². K]	1.8	6.12	6.12	6.12	6.12	6.12	6.12	6.12	6.12	6.12	6.12	6.12	6.12	6.12
UPS operation mode	[−]	0	0	2	1	1	1	1	1	1	1	1	1	1	1
UPSoversized power converter	[−]	0	0	0	0	1	1	1	1	1	1	1	1	1	1
WSP air containment	[−]	0	0	0	0	0	1	1	1	1	1	1	1	1	1
Variable flow rate	[−]	0	0	0	0	0	0	1	1	1	1	1	1	1	1
Max. Ambient temp.															
IAFC	[°C]	−100	−100	−100	−100	−100	−100	−100	AHU_Tsupply-20	AHU_Tsupply-5	AHU_Tsupply-5	AHU_Tsupply-5	AHU_Tsupply-5	AHU_Tsupply-5	AHU_Tsupply-5
WSP supply temp.	[°C]	20	20	20	20	20	20	20	20	24	24	24	24	24	24
Air temp. increase in WSP	[K]	7	7	7	7	7	7	7	7	7	15	15	15	15	15
VCCH nominal COP	[−]	4	4	4	4	4	4	4	4	4	4	5.5	5.5	5.5	5.5
Virtualization technique	[−]	0	0	0	0	0	0	0	0	0	0	0	1	1	1
Turn off idle server	[−]	0	0	0	0	0	0	0	0	0	0	0	0	1	1
IT scheduling	[−]	0	0	0	0	0	0	0	0	0	0	0	0	0	1

The non-renewable primary energy is influenced mainly by thermal containment, thermal indirect cooling, IT consolidation and turn off idle server. These measures lead to a significant decrease (46% for Barcelona and 48% for Stockholm) of the metric values. Due to a decreasing OPEX (33% for Barcelona and 30% for Stockholm) and slightly increasing CAPEX (9% for Barcelona and Stockholm) with the application of the measures the TCO shows a slight decrease of about 18% for Barcelona and 12% for Stockholm as well. The results of the PUE in both locations show the most significant decrease (22% for Barcelona and 26% for Stockholm) due to the

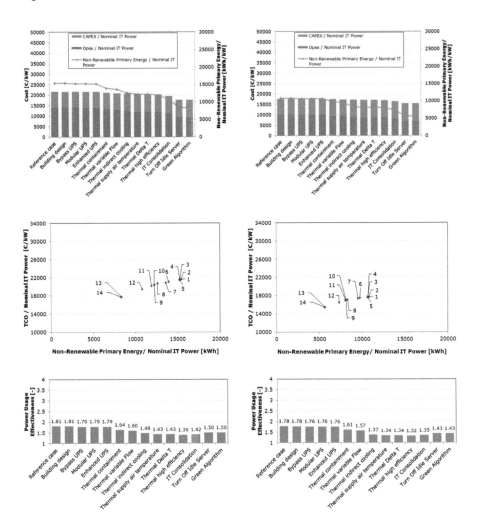

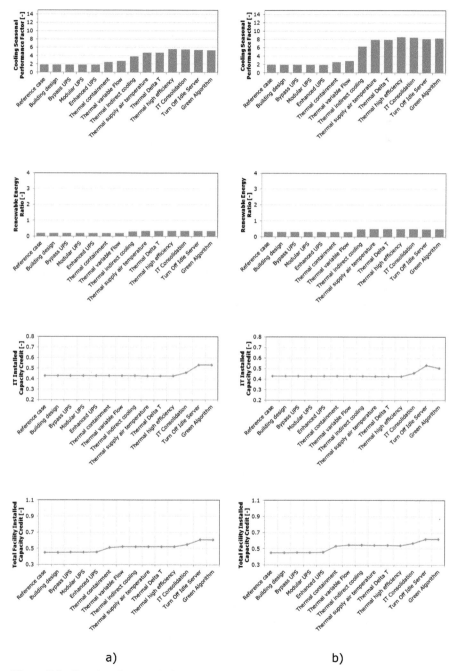

Figure 7.9 Results of accumulative tests for concept 1 in a) Barcelona and in b) Stockholm.

Table 7.6 Accumulation test results for the size of the main elements in concept 1

Components	Unit	Case													
		1	2	3	4	5	6	7	8	9	10	11	12	13	14
VCCh chiller (nominal power)	kW	1500	1500	1500	1500	1500	1500	1500	1500	1500	1500	1500	1500	1500	1500
Dry cooler (nominal power)	kW	1875	1875	1875	1875	1875	1875	1875	1875	1875	1875	1773	1773	1773	1773
Total air mass flow rate	t/h	768	768	768	768	768	302	302	302	302	302	302	302	302	302
Power consumption of fans	kW	94	94	94	94	94	37	37	37	37	37	37	37	37	37
Transformer (Nominal power)	kW	2169	2169	2169	2169	2169	2019	2019	2019	2019	2019	1879	1879	1879	1879
Switchgear (Nominal current of)	kW	4139	4139	4139	4139	4139	3853	3853	3853	3853	3853	3587	3587	3587	3587
Generator (Nominal power)	kW	2500	2500	2500	2500	2500	2500	2500	2500	2500	2500	2500	2500	2500	2500
Uninterruptible Power Supply unit (Nominal power)	kW	629	629	629	629	629	629	629	629	629	629	629	629	629	629
Power Distribution unit (Nominal power)	kW	2500	2500	2500	2500	2500	2500	2500	2500	2500	2500	2500	2500	2500	2500

use of thermal containment, thermal variable flow, thermal indirect cooling and the increase of supply air temperature. However a marginal increase of the metric value due to IT consolidation and turn off idle server is reached. The most noticeable increase (6% for Barcelona and 78% for Stockholm) of the cooling seasonal performance factor with the application of the efficiency measures can be found for thermal containment, thermal variable flow, thermal indirect cooling, the increase of supply air temperature and the use of high efficiency thermal elements whereas the application of IT consolidation and turn off idle servers lead to a slight decrease of the metric. This behaviour is valid for Barcelona and Stockholm as well. The RER increases significantly (39% for Barcelona and Stockholm) mainly due to the application of thermal indirect cooling for both locations equally. The IT installed capacity credit shows an independent behaviour towards the applied measures except for IT consolidation and turn off idle server. They lead to a significant increase (19% for Barcelona) of the metric values. This behaviour is equal for the two locations whereas for Stockholm the use of green algorithm leads to slightly decreasing values. Compared to the IT installed capacity credit, the total facility installed capacity credit shows similar dependencies but additionally presents increasing values with the use of thermal containment.

When analysing the sizes of the main elements the nominal power of the dry cooler, the total air mass flow rate, the power consumption of the fans, the nominal power of the transformer and the nominal current of the switchgear show a dependency on single applications of the efficiency measures. While the use of thermal containment leads to a reduction of the total air mass flow rate, the power consumption of the fans, the nominal power of the transformer and the nominal current of the switchgear the nominal power of the dry cooler is not affected. Also the application of high efficiency thermal elements components leads to decreasing values of the nominal power of the dry cooler, the nominal power of the transformer and the nominal current of the switchgear.

7.3.2.2 Influence of size

The variation of the Data Centre nominal IT power shown in Figure 7.10 is evaluated for the TCO and the RER in two locations Barcelona and Stockholm. Regarding the costs it can be analysed that the decrease with the application of all energy efficiency measures is more distinctly for Barcelona (15–20%) then for Stockholm (10–15%), however, the absolute cost values for Barcelona are 10–20% higher. The decrease of the non-renewable primary energy values with the application of the efficiency measures shows a similar tendency.

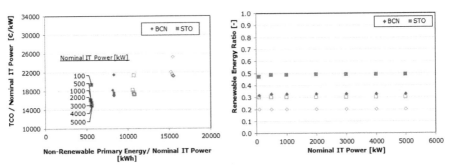

Figure 7.10   Variation of the Data Centre nominal IT power for the two locations (Barcelona and Stockholm) for concept 1. Unfilled shapes represent the reference case and filled shapes the case with all energy efficiency measures applied.

However, the decrease of about 47–48% is equal for both locations, whereas the absolute value is 30–32% higher for Barcelona than for Stockholm. The dependency of the nominal IT power is comparable for both locations and lead to decreasing costs of maximal 21% for Barcelona and 23% for Stockholm while the influence on the non-renewable primary energy is insignificant. The highest decrease of the costs can be analysed from 100 kW to 500 kW with a maximum cost reduction of 15% compared to the next variation steps (500, 1000, 2000 etc.) with a maximal decrease of 2%. Regarding the RER, the application of energy efficiency measures leads to a significant increase of maximal 39% for both locations. IT can be concluded that the location and neither the size of the Data Centre show a major dependency on this metric.

7.3.2.3 On-Site renewable energy systems implementation

Figure 7.11 depicts graphically the results of applying on-site renewable power systems to a conventional Data Centre: PV and wind turbines systems. On one hand, different sizes of on-site PV systems, which are characterized by their PV peak power, have been simulated. For the scenarios of 120 kW_{IT} and 400 kW_{IT}, PV systems are varied from 0 to 50 kWp and from 0 to 100 kWp, respectively (see Table 7.7). PV simulation which is integrated as part of the overall TRNSYS Data Centre models neglects the shadows effects of surrounding buildings and the own shadows of a large flat roof mounted PV field. The PV field is considered oriented south with an inclination slope of 32.2° for Barcelona and 44.6° for Stockholm. On the other hand, the use of

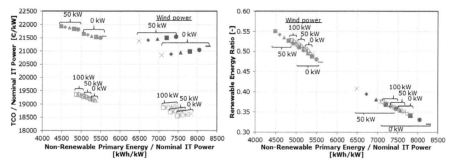

Figure 7.11 Normalized TCO (left) and RER (right) vs. normalized PE_{nren} for different scenarios when adding on-site renewable power systems to concept 1.

Table 7.7 Specific assumptions of on-site PV and wind power systems for the investigated concept 1

Parameter	Unit	BCN-120/STO-120	BCN-400/STO-400
Total PV peak power	[kWp]	0, 10, 20, 30, 40, 50	50, 60, 70, 80, 90, 100
Total Wind rated power	[kW]	0, 50	0, 50, 100

small wind power systems has also been studied. For the 120 kW$_{IT}$ Data Centre a unique 50 kW rated power wind turbine (Aeolos, 2015) is considered; while, for the 400 kW$_{IT}$ Data Centre, the impact of having one and two identical wind turbines has been calculated.

As expected, adding on-site PV and wind power systems implies an increase of the RER, as well as a decrease of the PE_{nren}. However, when analysing the economic and the energetic impact, the location of the Data Centre has a big influence. It is shown that on-site PV systems are cost-effective when they are implemented in Data Centres located in Barcelona (south Europe) under the hypothesis of this study but the installation of PV systems in Stockholm are not cost-effective due to investment, the electricity prices considered and the low solar radiation over the year. Renewable electricity produced by on-site wind power systems in Barcelona and in Stockholm, which is based on wind availability in Meteonorm data files [5], is not enough to compensate the investment needed for such a systems. Using on-site wind power systems needs to be installed in locations where wind resource is available, which strongly depends on local conditions. Table 7.8 shows the required roof space for a 50 kWp and 100 kWp PV field under the hypothesis that the PV field is mounted in a flat roof by tilted modules with a distance between rows based on rules which optimally minimize the occupancy of the

Table 7.8   Required roof space, available roof space and load cover factor for different scenarios with on-site PV power system for the investigated concept 1

Parameter	Unit	Name of the Scenarios			
		BCN-120	BCN-400	STO-120	STO-400
Required roof space – 50 kWp	[m²]	796		1570	
Required roof space – 100 kWp	[m²]	1593		3140	
Available roof space	[m²]	198	660	198	660
Load cover factor – 50 KWp	[%]	16	5	10	3
Load cover factor – 100 KWp	[%]	–	9	–	6

roof and maximizes the PV production [6]. The required roof space depends on the size of the PV system and on the location. Although different types of Data Centre buildings exist and whitespace rooms can be part of large corporate buildings, available roof space (A_{roof}) is estimated in relation with the white space area ($A_{DC,room}$) as shown in Equation (7.1) and based on information available from the industry. In some cases, to cover the expected PV peak power, The required roof space for the PV field exceeds the available space. Results of the load cover factor, which represents the ratio between the power produced by PV and the overall electrical Data Centre consumption, are also presented in Table 7.8 The load cover factor is very low (less or equal to 10% in most of the cases) even with PV systems that go beyond the available roof space. This means, that under grid parity conditions having an on-site PV system is a good solution to reduce the environmental impact of a Data Centre. Although this will be very dependent on the case, conventional roof mounted PV fields are limited by the available roof space covering a small portion of the electricity consumption of a Data Centre. Using PV to have larger load cover factors would require to use additional space available in the Data Centre footprint or explore building integrated PV solutions.

$$A_{roof} = f_{roof-ws} \cdot A_{DC,room};$$
$$\text{where } f_{roof-ws} \cdot = 2.2 m_{roof}^2 / m_{DC,room}^2 \qquad (7.1)$$

7.3.3 Concept 2. District Cooling and Heat Reuse

7.3.3.1 Influence of energy efficiency measures

For concept 2 the results of the accumulative tests are presented in Figure 7.12 (results for each metric analysed) and Table 7.9 (sizes of the main elements).

The energy efficiency measures influencing the results of the non-renewable primary energy are thermal containment, thermal indirect cooling, IT consolidation and turn off idle server. They lead to a significant decrease (42% for Barcelona and 47% for Stockholm) of the metric values. A decreasing OPEX (23% for Barcelona and 18% for Stockholm) and slightly increasing CAPEX (11% for Barcelona and Stockholm) with the application of all measures cause a slight decrease of the TCO of about 10% for Barcelona and 4% for Stockholm. The most significant decrease of 21% for Barcelona

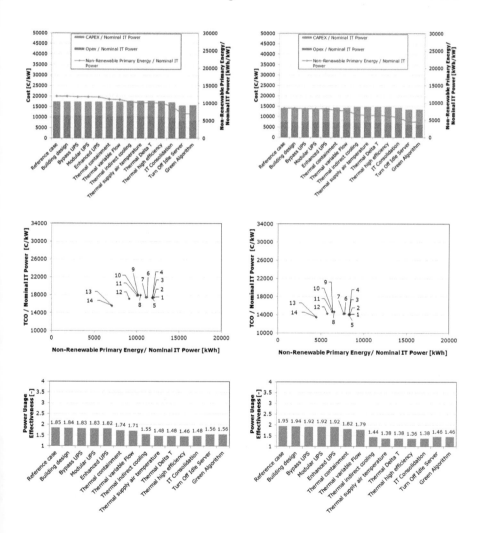

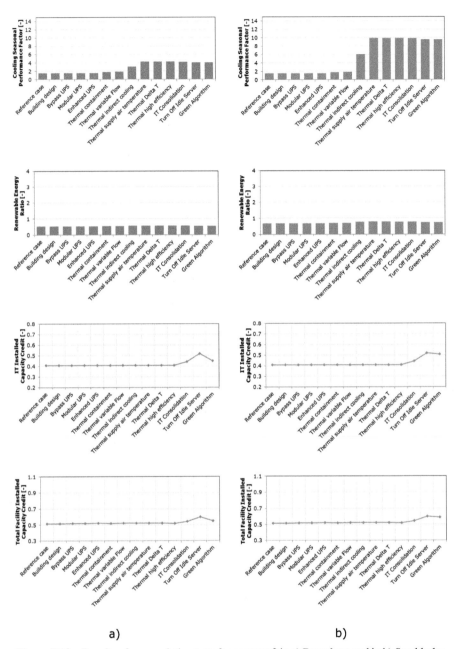

a)   b)

Figure 7.12 Results of accumulative tests for concept 2 in a) Barcelona and in b) Stockholm.

Table 7.9 Accumulation test results for the size of the main elements in concept 2

Components	Unit	\multicolumn{14}{c}{Case}
		1	2	3	4	5	6	7	8	9	10	11	12	13	14
Heat pump (heating capacity)	kW	175	175	175	175	175	175	175	175	175	175	175	175	175	175
Total air mass flow rate	t/h	544	544	544	544	544	214	214	214	214	214	214	214	214	214
Power consumption of fans	kW	67	67	67	67	67	26	26	26	26	26	26	26	26	26
Transformer (TR) (Nominal power)	kW	1758	1758	1758	1758	1758	1651	1651	1651	1651	1651	1589	1589	1589	1589
Switchgear (SWG) (Nominal current of)	kW	3354	3354	3354	3354	3354	3151	3151	3151	3151	3151	3033	3033	3033	3033
Generator (GE) (Nominal power)	kW	2500	2500	2500	2500	2500	2500	2500	2500	2500	2500	2500	2500	2500	2500
Uninterruptible Power Supply unit (UPS) (Nominal power)	kW	629	629	629	629	629	629	629	629	629	629	629	629	629	629
Power Distribution unit (PDU) (Nominal power)	kW	2500	2500	2500	2500	2500	2500	2500	2500	2500	2500	2500	2500	2500	2500

and 30% for Stockholm of the PUE can be found due to the use of thermal containment, thermal variable flow, thermal indirect cooling, the increase of supply air temperature and the use of high efficiency thermal elements. However, a slight increase of the metric value due to IT consolidation and turn off idle server is reached. The cooling seasonal performance factor increases significantly (65% for Barcelona) with the application of the efficiency measures such as thermal containment, thermal variable flow, thermal indirect cooling and the increase of supply air temperature whereas the application of IT consolidation and turn off idle server lead to a slight decrease of the metric. This behaviour is valid for the two locations equally. However, for Stockholm the increase of 85% of the cooling seasonal performance factor is mainly due to the efficiency measure thermal indirect cooling and the increase of supply air temperature. Regarding the RER an increase of 9% for Barcelona and 13% for Stockholm is mainly caused due to the application of thermal indirect cooling and the increase of supply air temperature, while IT consolidation and turn off idle server lead to an insignificant decrease. The IT installed capacity credit shows no dependency towards the applied measures except for IT consolidation, turn off idle server and green algorithm. While the first two measures lead to a significant increase of 23% for Barcelona and for Stockholm the use of green algorithm leads to decreasing values for Barcelona (13%) and Stockholm (4%) as well. Compared to the IT installed capacity credit, the total facility installed capacity credit shows similar dependencies just in a less distinctive way.

When analysing the sizes of the main elements the total air mass flow rate, the power consumption of the fans, the nominal power of the transformer and the nominal current of the switchgear show a dependency on single applications of the efficiency measures. These are comparable to the analysis done for concept 1 (see Subsection 7.3.2.1).

7.3.3.2 Influence of size

The variation of the Data Centre nominal IT power shown in Figure 7.13 is evaluated for the TCO and the RER. Analyzing the effects in costs of the energy efficiency strategies implementation, it is seen that locations such as Barcelona (8–12%) is more influenced by this than Stockholm (2–5%). However, the absolute cost values for Barcelona are 12–20% higher. Regarding the non-renewable primary energy values a decrease of 42% can be reached for Barcelona and 47% for Stockholm. The dependency of the nominal IT power is comparable for both locations and leads to decreasing costs up to 18% for Barcelona and 20% for Stockholm while the influence on the non-renewable

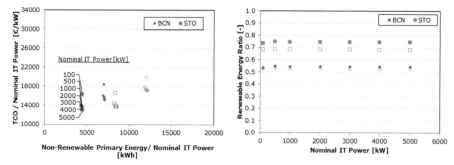

Figure 7.13 Variation of the Data Centre nominal IT power for the two locations (Barcelona and Stockholm) for concept 2. Unfilled shapes represent the reference case and filled shapes the case with all energy efficiency measures applied.

primary energy is insignificant. The most notable decrease of the costs can be seen from 100 kW to 500 kW with a maximum cost reduction of 14% compared to the next variation steps (500 kW, 1000 kW, 2000 kW, etc.) with a maximum decrease of 3%. Regarding the RER, the application of energy efficiency measures lead to a significant increase up to 4% for Barcelona and 8% for Stockholm. The size of the Data Centre does not show a major dependency of this metric.

7.3.3.3 Influence of the liquid cooling solution and the potential heat reuse

Figure 7.14 presents the results of the parametric analysis for concept 2 (Data Centre connected to a district cooling and heating system with reuse of heat from direct liquid cooled servers) for Data Centres of 400 kW$_{IT}$ in both locations (Barcelona-400 and Stockholm-400). Only results for 400 kW$_{IT}$ are presented to contribute to readability of the graphs. Maintaining the hypothesis that 100% of the heat extracted from the Data Centre can be reused, the sensitivity analysis for two parameters has been performed. On one hand, there is the type of liquid cooling system which is characterized by its efficiency: 0.65 for on-chip liquid cooling and 1.0~0.99 for immersed liquid cooling system. This parameter means that for on-chip liquid cooling, 65% of the heat is extracted by water and the other 35% by air, while for immersed liquid cooling systems, 100% of the heat is extracted by water. On the other hand, the size of the heat pump is varied which is characterized by the ratio between the heat pump cooling power and the maximum liquid cooling demand of the Data Centre. This ratio is varied from 0.1 to 0.5. The results

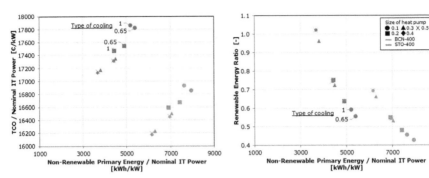

Figure 7.14 Normalized TCO (left) and RER (right) vs. normalized PE_{nren} for different scenarios, types of cooling (On-chip/Immersed) and sizes of the absorption chiller for concept 2.

show that as the type of liquid cooling system allows extracting higher amount of heat from the servers, PE_{nren} decreases as well as RER increases. As the ratio determining the size of the heat pump increases there is a reduction of the PE_{nren}, too. Having an immersed liquid cooling system with the capability to extract more heat is a bit more expensive, but with optimum sizes of the heat pump differences between liquid cooling technologies are minimized in terms of TCO and PE_{nren}. As it is shown in Figure 7.14, the size of the heat pump has an important effect on the indicators having an optimal value between 0.4 and 0.5, while the differences between these two values of the parameter are negligible.

7.3.4 Concept 3. Grid-Fed Wet Cooling Tower without Chiller

7.3.4.1 Influence of energy efficiency measures
The results of the accumulative tests for concept 3 are presented in Figure 7.15 (results for each metric analysed) and Table 7.10 (the sizes of the main elements of each concept).

Concerning the non-renewable primary energy the highest dependency of the metrics can be analysed with the application of thermal containment, IT consolidation and turn off idle server in both locations. They lead to a significant decrease of 40% for Barcelona and Stockholm. The same efficiency measures lead to a decreasing OPEX (28% for Barcelona) and slightly increasing CAPEX (4% for Barcelona) which causes a slight decrease of

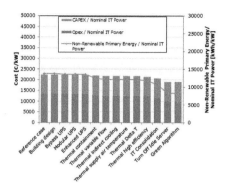
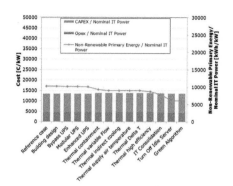

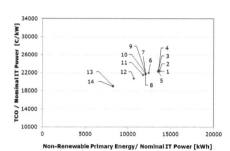
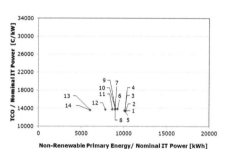

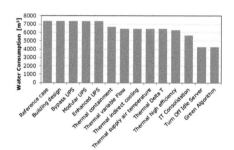
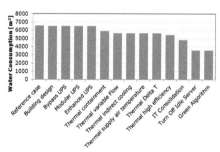

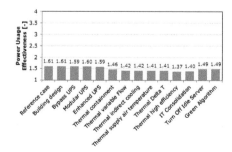
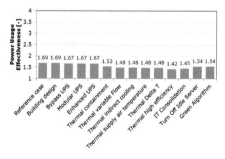

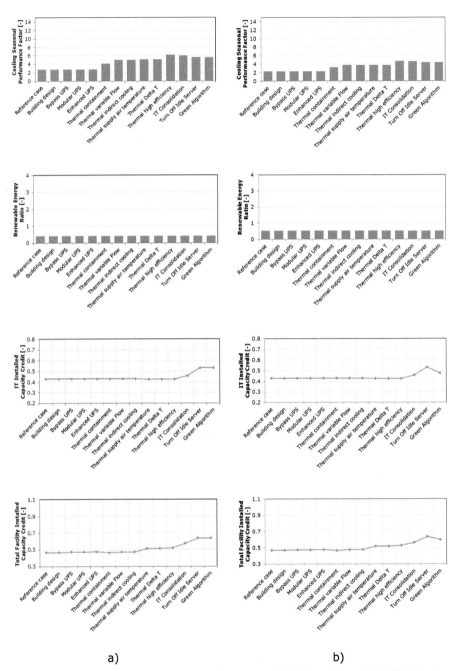

a)   b)

Figure 7.15   Results of accumulative tests for concept 3 in a) Barcelona and in b) Stockholm.

the TCO of about 15% for Barcelona as well. For the location Stockholm the increase of the CAPEX and the decrease of the OPEX compensate and lead to nearly constant TCO values. The water consumption is analysed due to the fact that in concept 3 a wet cooling tower is applied. Decreasing values of this metric can be observed due to the application of thermal containment and thermal variable flow as well as IT consolidation and turn off idle server. The overall reduction of the water consumption applying all energy efficiency measures reached up to 42% for Barcelona and 46% for Stockholm. The results of the PUE in both locations show the most significant decrease due to the use of thermal containment but also thermal variable flow and the use of high efficiency thermal elements. However, a slight increase of the metric value due to IT consolidation and turn off idle server is reached. Regarding the cooling of the seasonal performance factor the most noticeable increase is shown with the application of thermal containment, thermal variable flow and the use of high efficiency thermal elements. However, the efficiency measures IT consolidation and turn off idle server lead to a slight decrease of the metric in both locations equally. The decrease of the RER with all efficiency measures applied is marginal with 2% in both locations. The IT installed capacity credit shows a dependency on IT consolidation and turn off idle server. These measures lead to a significant increase of the metric values. This behaviour is equal for the two locations whereas for Stockholm the use of green algorithm leads to decreasing values. The results of the total facility installed capacity credit show very similar tendencies except for a further dependency on the increase of supply air temperature.

When analysing the sizes of the main elements, the effect of the efficiency measures on the total air mass flow rate, the power consumption of the fans, the nominal power of the transformer and the nominal current of the switchgear is very similar to the results for concept 1 (see Subsection 7.3.2.1). Additionally, the specific value for concept 3 (cooling tower) shows a decreasing dependency on the measure using high efficiency thermal elements.

7.3.4.2 Influence of EE measures

Table 7.10 Accumulation test results for the size of the main elements in concept 3

Components	Unit	Case 1	2	3	4	5	6	7	8	9	10	11	12	13	14
VCCh chiller (nominal power)	kW	1500	1500	1500	1500	1500	1500	1500	1500	1500	1500	1500	1500	1500	1500
Cooling tower (nominal power)	kW	1875	1875	1875	1875	1875	1875	1875	1875	1875	1875	1773	1773	1773	1773
Total air mass flow rate	t/h	768	768	768	768	768	302	302	302	302	302	302	302	302	302
Power consumption of fans	kW	94	94	94	94	94	37	37	37	37	37	37	37	37	37
Transformer (TR) (Nominal power)	kW	2185	2185	2185	2185	2185	2034	2034	2034	2034	2034	1899	1899	1899	1899
Switchgear (SWG) (Nominal current of)	kW	4169	4169	4169	4169	4169	3883	3883	3883	3883	3883	3623	3623	3623	3623
Generator (GE) (Nominal power)	kW	2500	2500	2500	2500	2500	2500	2500	2500	2500	2500	2500	2500	2500	2500
Uninterruptible Power Supply unit (UPS) (Nominal power)	kW	629	629	629	629	629	629	629	629	629	629	629	629	629	629

7.3.4.3 Influence of size

The variation of the Data Centre nominal IT power shown in Figure 7.16 is evaluated for the TCO and the RER in two locations Barcelona and Stockholm. Regarding the costs it can be analysed that the decrease with the application of all energy efficiency measures for Barcelona is 13–16% while the values for Stockholm are marginally increasing (1–2%). However, the cost values for Barcelona, are 23–41% higher than for Stockholm. The decrease of the non-renewable primary energy values with the application of the efficiency measures reaches up to 40–42%. The dependency of the nominal IT power is comparable for both locations and lead to decreasing costs of maximal 20% for Barcelona and 27% for Stockholm while the influence on the non-renewable primary energy is insignificant. The most significant decrease of the costs can be analysed from 100 kW to 500 kW compared to the next variation steps (500 kW, 1000 kW, 2000 kW, etc.). Regarding the RER, the application of energy efficiency measures lead to a slight increase of about 5% for Barcelona and 4% for Stockholm, while the size of the Data Centre does not show a major dependency on this metric.

7.3.4.4 On-site PV systems implementation

Figure 7.17 shows the results of applying on-site PV power systems to concept 3 for the different scenarios. Results are in coherence with the findings of adding PV to concept 1. PE_{nren} decreases and RER increases when the size of the PV system increases, being cost-effective in Barcelona and not economically feasible for Stockholm in terms of TCO.

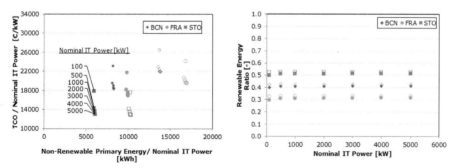

Figure 7.16 Variation of the Data Centre nominal IT power for the two locations (Barcelona and Stockholm) for concept 3. Unfilled shapes represent the reference case and filled shapes the case with all energy efficiency measures applied.

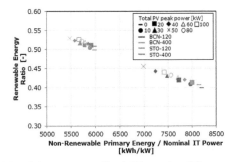

Figure 7.17 Normalized TCO (left) and RER (right) vs. normalized PE$_{nren}$ for different scenarios when adding on-site PV power systems to concept 3.

7.3.5 Concept 4. Grid-Fed Vapour-Compression Chiller with Electrical Energy and ChilledWater Storages

7.3.5.1 Influence of EE measures

The results for each metric analysed (Figure 7.18) and the sizes of the main elements (Table 7.11) present the results of the accumulative tests for concept 4.

A significant decrease (40% for Barcelona and 41% for Stockholm) of the non-renewable primary energy values is influenced mainly by thermal containment, thermal indirect cooling, IT consolidation and turn off idle server in both locations. Due to a decreasing OPEX (25% for Barcelona) and slightly increasing CAPEX (8% for Barcelona) with the application of the measures the TCO shows a slight decrease of about 13% for Barcelona as well. For the location Stockholm the increase of the CAPEX (8%) and the decrease of the OPEX (4%) lead to a slight increase of the TCO values of about 4%. The water consumption decreases due to the application of thermal containment, thermal indirect cooling, the increase of supply air temperature, IT consolidation and turn off idle server. In both locations the highest dependency is seen for thermal indirect cooling. The overall reduction of the water consumption applying all energy efficiency measures reaches up to 75% for Barcelona and 90% for Stockholm. The results of the PUE in both locations show a slight decrease due to the use of thermal containment, thermal variable flow, thermal indirect cooling, the increase of supply air temperature and high efficiency thermal elements while IT consolidation and turn off idle server lead to slightly increasing values of the PUE. The most noticeable increase (up to 67% for Barcelona and 76% for Stockholm) of the cooling seasonal performance factor with the application of the efficiency measures can be found for thermal containment, thermal variable flow, thermal

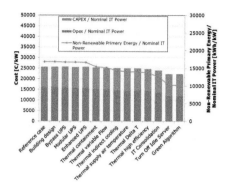
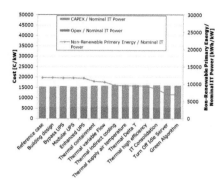

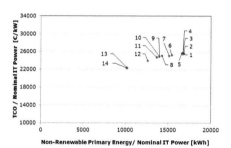
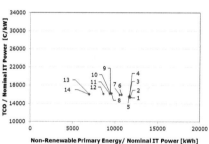

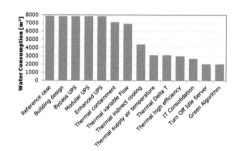
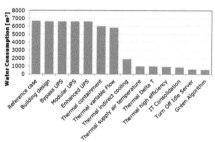

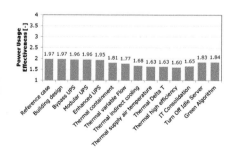
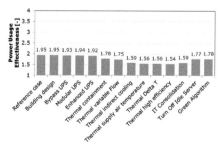

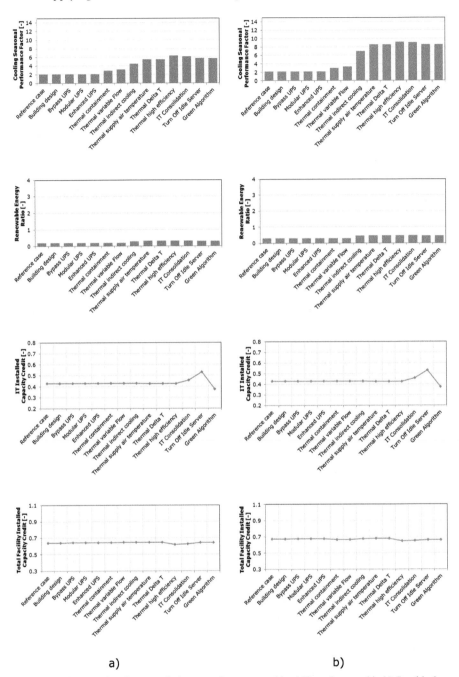

a)   b)

Figure 7.18   Results of accumulative tests for concept 4 in a) Barcelona and in b) Stockholm.

Table 7.11 Accumulation test results for the size of the main elements in concept 4

Components	Unit	Case													
		1	2	3	4	5	6	7	8	9	10	11	12	13	14
VCCh chiller (nominal power)	kW	1500	1500	1500	1500	1500	1500	1500	1500	1500	1500	1500	1500	1500	1500
Volume of the CHWST	m³	1237	1237	1237	1237	1237	1237	1237	1237	1237	1237	1237	1237	1237	1237
Total air mass flow rate	t/h	768	768	768	768	768	302	302	302	302	302	302	302	302	302
Power consumption of fans	kW	94	94	94	94	94	37	37	37	37	37	37	37	37	37
Transformer (TR) (Nominal power)	kW	2537	2537	2537	2537	2537	2386	2386	2386	2386	2386	2142	2142	2142	2142
Switchgear (SWG) (Nominal current of)	kW	6560	6560	6560	6560	6560	6172	6172	6172	6172	6172	5539	5539	5539	5539
Generator (GE) (Nominal power)	kW	2500	2500	2500	2500	2500	2500	2500	2500	2500	2500	2500	2500	2500	2500
Uninterruptible Power Supply unit (UPS) (Nominal power)	kW	629	629	629	629	629	629	629	629	629	629	629	629	629	629
Power Distribution unit (PDU) (Nominal power)	kW	2500	2500	2500	2500	2500	2500	2500	2500	2500	2500	2500	2500	2500	2500

indirect cooling, the increase of supply air temperature and the use of high efficiency thermal elements, whereas the application of IT consolidation and turn off idle servers lead to a slight decrease of the metric. This behaviour is valid for Barcelona and Stockholm as well. The RER increases significantly (61% for Barcelona and 62% for Stockholm) with all efficiency measures applied, mainly induced by the application of thermal indirect cooling and turn off idle server. The IT installed capacity credit shows an independent behaviour towards the applied measures except for IT consolidation, turn off idle server and green algorithm. The first two lead to a significant increase of the metric values (19%), while the use of green algorithm leads to decreasing values that end nearly at the reference value in both locations equally. The results of the total facility installed capacity credit show nearly no dependency on measures applied but except for a slight increase with the application of thermal containment, the use of high efficiency thermal elements and a slight increase with the applications of IT consolidation and turn off idle server. This analysis is equally valid for both locations.

When analysing the sizes of the main elements, the effect of the efficiency measures on the total air mass flow rate, the power consumption of the fans, the nominal power of the transformer and the nominal current of the switchgear show similar dependencies compared to the results for concept 1 (see Subsection 7.2.1).

7.3.5.2 Influence of size

In Figure 7.19 the variation of the Data Centre nominal IT power is shown for the TCO and the RER. Regarding the results of the TCO in the two locations Barcelona and Stockholm a notable decrease with the application of all energy efficiency measures is shown for Barcelona (9–14%) while for

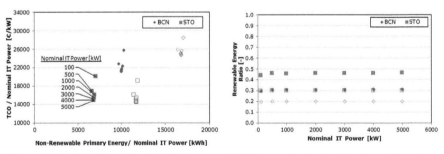

Figure 7.19 Variation of the Data Centre nominal IT power for the two locations (Barcelona and Stockholm) for concept 4. Unfilled shapes represent the reference case and filled shapes the case with all energy efficiency measures applied.

Stockholm a marginal increase can be analysed (3–5%). The absolute cost values for Barcelona are 22–42% higher than for Stockholm. The decrease of the non-renewable primary energy values with the application of the efficiency measures reaches up to 39–42% nearly equally in both locations. The dependency of the nominal IT power is comparable for both locations and lead to decreasing costs of maximal 18% for Barcelona and 26% for Stockholm while the influence on the non-renewable primary energy is nearly insignificant. The most significant decrease of the costs can be analysed from 100 kW to 500 kW compared to the next variation steps (500, 1000, 2000 etc.). When looking at the RER, the implementation of the energy efficiency strategies lead to a notable increase of about 33% for Barcelona and 35% for Stockholm, while the size of the Data Centre does not show a major dependency on this metric.

7.3.5.3 Influence of the size of TES

It is well known that the use of energy storage systems can play an important role to reduce operational costs of any infrastructure where there are differences between energy production and energy demand or between electricity prices over the day (i.e. day/night price). The main drawback of implementing TES systems in real applications is first, the investment cost which needs to be carefully analysed, and second, each TES implementation needs a careful energy analysis considering the operational boundary conditions. It is reasonable to expect that an optimal configuration, including energy storage and energy management, exists for a current situation and in particular for a specific Data Centre characteristics. For this reason, it is highly recommended to develop energy and economic dynamic models and optimize the system configuration with the real boundary conditions.

Standard Data Centres have redundant chiller which can be used to produce chilled water when the electricity cost is low during off-peak hours and store that cold in the TES system. Contrarily, when the electricity price is expensive (peak hours), the management system will enforce the tank to be discharged in order to decrease the return water temperature. The optimal configuration of the TES implementation concerns combinations of choices regarding the storage tank volume and the operational performance of the chiller used to produce cold water during off-peak hours in terms of desired chilled water temperature and return water temperature to switch on the chiller.

In order to show the methodology for TES implementation, an operative Data Centre with a total IT capacity of 140 kW is considered. This facility is currently being used to provide computing and information services for the

Polytechnic University of Catalonia, in Barcelona (Spain). The IT consumption of the infrastructure (70 racks of data and 12 racks of communication equipment) is 100 kW IT while the IT room area is of 285 m^2.

In the optimization phase the TCO, which takes also into consideration the investment cost of the TES system implemented, should be minimized. The economic objective function considers two different of expenses: the operational expenditures (OPEX) and the investment cost (CAPEX) as shown in Equation (7. 2).

$$TCO = OPEX + CAPEX \tag{7.2}$$

where OPEX and CAPEX are calculated using Equation (7.3) and Equation (7.4), respectively.

$$OPEX = \left(\sum_{i=1}^{3} E_{chiller}^{i} + \sum_{i=1}^{3} E_{pump}^{i} + E_{CRAH} \right) \cdot p \tag{7.3}$$

where $E_{chiller}$ is the energy consumption of the chillers in kWh, E_{pump} is the energy consumption of the pumps in kWh, E_{CRAH} is the energy consumption of the CRAHs in kWh, and p is the electricity price in €/kWh.

$$CAPEX = (tankCost + waterCost) \cdot V_{TES} \tag{7.4}$$

where tankCost is the cost for the storage tank as well as all the auxiliary equipment necessary in € and waterCost is the water cost.

Table 7.12 shows the decision variables of the problem as well as the theoretical upper/lower bound for the optimization process for the real Data Centre analysed. Regarding the upper/lower bound for the water storage tank volume, it makes no sense to study systems smaller than 5 m^3 since the impact will be really poor. On the other hand, the actual chiller capacity (77.7 kW$_{th}$) limits the storage at 150 m^3. Above this storage volume, the chiller is not able to charge the tank completely.

Table 7.12   Description of the decision variables

Decision Variable	Units	Range
Water storage tank volume	[m^3]	[5,150]
Chiller outlet water temperature set-point	[°C]	[6,12.5]
Chiller temperature difference (inlet-outlet)	[°C]	[1,5]

Table 7.13 Optimization results for both scenarios

Scenario	$T_{sp,3}$ [°C]	V_{TES} [m³]	dT [°C]	Investment [€]	Operational Expenses [€]
Scenario 1 (RenewIT)	6	64.4	2.9	55,584	141,834
Scenario 2 (Fragaki)	6	47.8	1	42,851	142,491

Table 7.13 shows the yearly operational expenses for the upper/lower bound scenarios in comparison to the reference case, in case when no TES system is incorporated into the Data Centre and with the two optimized scenarios. RenewIT and Fragaki [7] scenarios represent similar investment cost of the TES system with different lifetime period. The results show that the OPEX values for upper/lower scenarios are lower than the reference system, so the strategy of storing cold during off-peak hours is proven but the investment cost should be taken into account. For instance, the mixed bound 1 scenario provides more annual savings than Fragaki but the investment cost is much higher; therefore, it is not recommended. Table 7.13 shows the results of the optimization, presenting the optimized operational values ($T_{sp,3}$ and dT), the storage tank volume (V_{TES}) and the associated investment and operational expenses. As expected, the investment function influences drastically the final results. On one hand, in scenario 1 (RenewIT) where the investment expenses are cheaper while the lifetime is higher (25 years), the optimal configuration is for a storage tank of 64.4 m³. On the other hand, in scenario 2 (Fragaki), where the lifetime of the system is being reduced to 15 years, the optimal volume tank is smaller (47.8 m³). The storage volume then affects the working temperatures of the chiller. In both cases, the desired chilled water temperature from the chiller ($T_{sp,3}$) is the lower bound (6°C). This phenomenon clearly opens the door to explore other systems which allow producing water at lower temperatures by the use of some refrigerant. Another strategy will be to use ice storage tanks. However, depending on the scenario analysed, the activation temperature is different. While for scenario 1, dT is 2.9°C and therefore $T_{ACT,3}$ is 8.9°C, for scenario 2 dT is the lower bound, with $T_{ACT,3}$ of 7°C.

Table 7.14 summarizes all the economic figures already described for both scenarios. These economic figures demonstrate the feasibility of the implementation of TES into Data Centre portfolio. However, the economic benefit after the lifetime of the system is not encouraging while in the future the cost of the electricity may change and therefore it is difficult to conclude that the investment for the TES system implementation, in particular with scenario 1, is beneficial for the Data Centre presented. For this reason, the impact of the uncertainty of the electricity price in the future is also studied.

Table 7.14 Economical figures for both scenarios analysed

Scenario	Investment Cost [€]	Yearly Energy Cost [€/y]	Annual Savings [€/y]	NPV [€]	Payback [years]	BCR [–]
Scenario 1 (RenewIT)	55,584	144,058	4,593	53,373	<12	1.52
Scenario 2 (Fragaki)	42,851	145,348	3,936	16,739	<11	1.15

When thermal or electric energy storage is implemented in any facility, the electricity price difference between peak and off-peak is more important than the absolute value of the electricity. Two scenarios varying the electricity price are evaluated considering that the average electricity price is the same (0.11 €/kWh) for each scenario while the amplitude between peak and off-peak hours is modified. Scenario 1 assumes an amplitude of 0.06 €/kWh and scenario 2 0.12 €/kWh. In this context, the implementation of a TES system is studied considering a storage tank of 50 m^3 and the operational conditions as $T_{sp,3}$ = 6°C and dT = 2°C. Figure 7.20 and Figure 7.21 show the yearly operational expenses for different systems: a Data Centre with no TES system (no TES), a Data Centre with TES and the actual electricity price (reference) and the same Data Centre with different electricity prices (scenario 1 and scenario 2). The results clearly show the importance of the

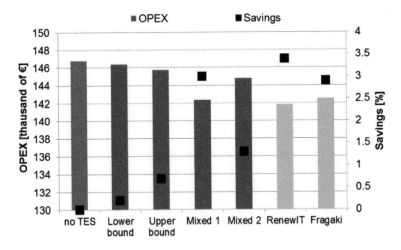

Figure 7.20 Yearly operational expenses for different scenarios.

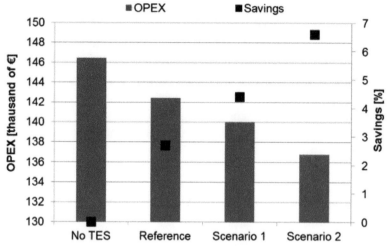

Figure 7.21 Yearly operational expenses for different electricity scenarios.

difference between peak and off-peak electricity price. For scenario 2, where this difference is higher, the operational savings can increase up to 7%. This results demonstrate that the use of TES is preferable in such countries where the electricity variation between day and night is important like in the USA where the difference can be higher than 0.08 €/kWh.

7.3.6 Concept 5. Biogas Fuel Cell with Absorption Chiller

7.3.6.1 Influence of EE measures

The results for each metric analysed (Figure 7.22) and the sizes of the main elements (Table 7.15) present the results of the accumulative tests for concept 5.

The non-renewable primary energy is influenced mainly by thermal containment, thermal indirect cooling, IT consolidation and turn off idle server in both locations analysed. These measures lead to a significant decrease (64% for Barcelona and 63% for Stockholm) of the metric values. Due to a decreasing OPEX (32% for Barcelona) and slightly increasing CAPEX (8% for Barcelona) with the application of the measures the TCO shows a slight decrease of about 17% for Barcelona as well. For Stockholm the increase of the CAPEX (8%) and the decrease of the OPEX (6%) lead to a marginal increase of the TCO values of about 1%. The results of the PUE in both locations decrease (12% for Barcelona and 14% for Stockholm) due to the use of thermal containment, thermal variable flow, thermal indirect cooling, the increase of supply air temperature and high efficiency thermal elements while the application of IT consolidation and turn off idle server lead to notable increasing values (26% for Barcelona and Stockholm) of the PUE. The significant increase (up to 66% for Barcelona and 82% for Stockholm) of the cooling seasonal performance factor with the application of the efficiency measures is mainly caused by thermal indirect cooling, the increase of supply air temperature but also thermal containment, thermal variable flow, and the use of high efficiency thermal elements contribute to this increase, whereas the application of IT consolidation and turn off idle servers lead to a slight decrease

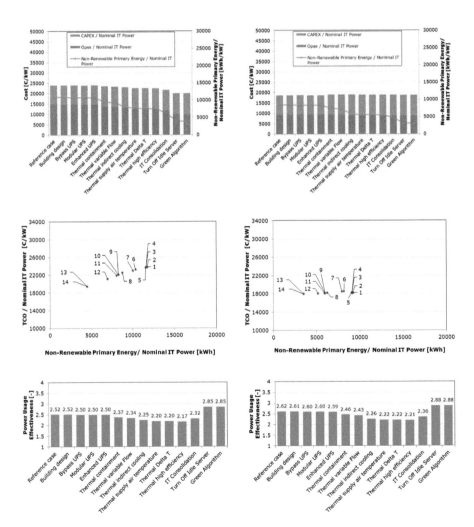

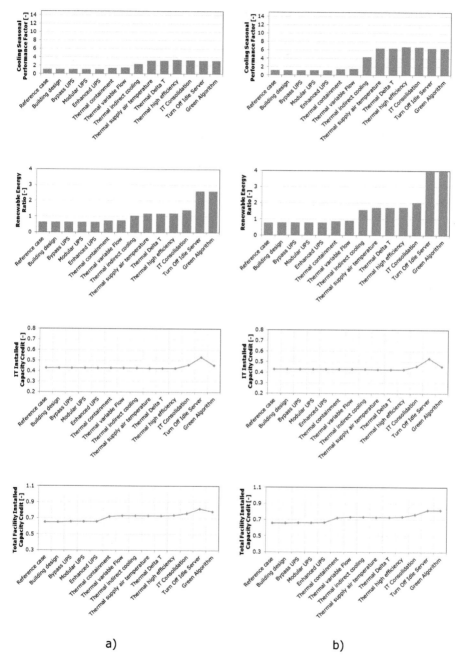

Figure 7.22 Results of accumulative tests for concept 5 in a) Barcelona and in b) Stockholm.

Table 7.15 Accumulation test results for the size of the main elements in concept 5

Components	Unit	Case													
		1	2	3	4	5	6	7	8	9	10	11	12	13	14
Absorption chiller (nominal power)	kW	288	288	288	288	288	288	288	288	288	288	288	288	288	288
Cooling tower (nominal power)	kW	634	634	634	634	634	634	634	634	634	634	634	634	634	634
Total air mass flow rate	t/h	768	768	768	768	768	302	302	302	302	302	302	302	302	302
Power consumption of fans	kW	94	94	94	94	94	37	37	37	37	37	37	37	37	37
Transformer (TR) (Nominal power)	kW	1873	1873	1873	1873	1873	1722	1722	1722	1722	1722	1647	1647	1647	1647
Switchgear (SWG) (Nominal current of)	kW	3574	3574	3574	3574	3574	3287	3287	3287	3287	3287	3144	3144	3144	3144
Generator (GE) (Nominal power)	kW	2500	2500	2500	2500	2500	2500	2500	2500	2500	2500	2500	2500	2500	2500
Uninterruptible Power Supply unit (UPS) (Nominal power)	kW	629	629	629	629	629	629	629	629	629	629	629	629	629	629
Power Distribution unit (PDU) (Nominal power)	kW	2500	2500	2500	2500	2500	2500	2500	2500	2500	2500	2500	2500	2500	2500

of the metric. This behaviour is valid for Barcelona and Stockholm as well. Regarding the RER an increase of 61% for Barcelona and 35% for Stockholm can be analysed with all efficiency measures applied, mainly induced by the application of thermal indirect cooling. The IT installed capacity credit shows an independent behaviour towards the applied measures except for IT consolidation, turn off idle server and green algorithm. The first two lead to a significant increase of the metric values, while the use of green algorithm leads to decreasing values that end up 11% below the reference value in both locations equally. The results of the total facility installed capacity credit show nearly no dependency on measures applied but except for a slight increase (about 20% for Barcelona and Stockholm) with the application of thermal containment, high efficiency thermal elements, IT consolidation and turn off idle server. For Barcelona a slight decrease can be found for the application of green algorithm. Other than that the analysis is equally valid for both locations.

When analysing the sizes of the main elements, the effect of the efficiency measures on the total air mass flow rate, the power consumption of the fans, the nominal power of the transformer and the nominal current of the switchgear show similar dependencies compared to the results for concept 1 (see Subsection 7.3.2.1).

7.3.6.2 Influence of size

Figure 7.23 shows the variation of the Data Centre nominal IT power for concept 5. The evaluation is done for the TCO and the RER in the two locations Barcelona and Stockholm. Regarding the costs it can be analysed that a notable decrease with the application of all energy efficiency measures is shown for Barcelona (12–18%) while for Stockholm a marginal increase can

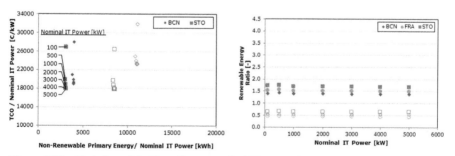

Figure 7.23 Variation of the Data Centre nominal IT power for the two locations (Barcelona and Stockholm) for concept 5. Unfilled shapes represent the reference case and filled shapes the case with all energy efficiency measures applied.

be analysed (0.2–1.6%). However, the absolute cost values for Barcelona are 4–23% higher. The decrease of the non-renewable primary energy values with the application of the efficiency measures reaches up to 64% equally in both locations. The dependency of the nominal IT power is comparable for both locations and lead to decreasing costs of maximal 32% for Barcelona and 34% for Stockholm while the influence on the non-renewable primary energy is nearly insignificant. The most significant decrease of the costs can be analysed from 100 kW to 500 kW (up to 21% for Barcelona and 26% for Stockholm) compared to the next variation steps (500 kW, 1000 kW, 2000 kW, etc.). Regarding the RER the application of energy efficiency measures lead to a notable increase of maximal 62% for Barcelona and Stockholm equally. However the size of the Data Centre does not show a notable dependency on this metric.

7.3.6.3 Influence of absorption chiller sizes and potential heat reuse

Figure 7.24 presents the results of a parametric analysis for concept 5 (Data Centre based on biogas fuel-cell driving an absorption chiller) for different scenarios. Figure 7.18 (right) only shows results for 400 kW$_{IT}$ Data Centre to improve the readability. The variation of two parameters has been analysed. One is the amount of heat produced by the facility which can be reused for other purposes outside the Data Centre: this is characterized by a ratio between 0.2 and 1.0, meaning 1.0 that 100% of the heat can be reused outside the Data Centre. The other parameter is the size of the absorption chiller which is characterized by a parameter from 0.2 to 0.5 which is the ratio between

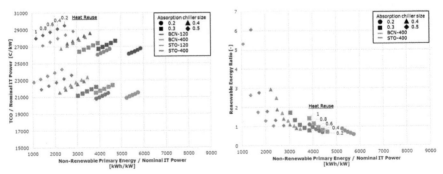

Figure 7.24   Normalized TCO (left) and RER (right) vs. normalized PE$_{nren}$ for different scenarios, relative absorption chiller sizes and amount of heat reuse ratio for concept 5.

absorption cooling capacity and the total cooling power of the Data Centre. The results in Figure 7.18 show that an increase of the absorption chiller size is not a cost-effective measure although it has an important effect on the reduction of the PE_{nren} and the increase of RER. For a ratio of the absorption chiller greater than 0.3, RER values are close to or great than 1.0 which indicates that the Data Centre facility is becoming a positive producer of primary energy even with low ratios of heat reuse. When the potential heat reuse is being reduced, the TCO increases as well as the PE_{nren} as consequence of the reduction of the exported heat.

7.3.7 Concept 6. Reciprocating Engine CHP with Absorption Chiller

7.3.7.1 Influence of EE measures

The results for each metric analysed (Figure 7.25) and the sizes of the main elements (Table 7.16) present the results of the accumulative tests for concept 6.

The results of the non-renewable primary energy are influenced mainly by thermal containment, thermal indirect cooling, IT consolidation and turn off idle server in both locations. These measures lead to a significant decrease (62% for Barcelona and Stockholm) of the metric values. Due to a decreasing OPEX (32% for Barcelona) and slightly increasing CAPEX (9% for Barcelona) with the application of the measures the TCO shows a slight decrease of about 18% for Barcelona as well. For the location Stockholm the increase of the CAPEX (9%) and the decrease of the OPEX (10%) lead to a slight decrease of the TCO values of about 1%. Regarding the results of the PUE in both locations a slight decrease of 13% for Barcelona and 15% for Stockholm can be analysed due to the use of thermal containment, thermal variable flow, thermal indirect cooling, the increase of supply air temperature and high efficiency thermal elements while IT consolidation and turn off idle server lead to increasing PUE values (23% for Barcelona and Stockholm) that reach even higher values than the reference case (11% for Barcelona and 9% for Stockholm). The significant increase (up to 66% for Barcelona and 82% for Stockholm) of the cooling seasonal performance factor with the application of all efficiency measures is caused mainly by thermal indirect cooling and the increase of supply air temperature but also thermal containment, thermal variable flow, and the use of high efficiency thermal elements contribute to this increase, whereas the application of IT consolidation and turn off idle servers lead to a marginal decrease of the metric. This behaviour is valid for Barcelona and Stockholm as well. For the RER an exceptional high increase of 75% for Barcelona and 80% for Stockholm with all efficiency measures applied can be analysed. This is mainly induced by the application of thermal indirect cooling, the increase of supply air temperature, IT consolidation and turn off idle server. Concerning the IT installed capacity credit the applied measures show an independent behaviour towards the applied measures except for IT consolidation, turn off idle server and green algorithm. While the first two lead to a significant increase of the metric values, the use of green algorithm

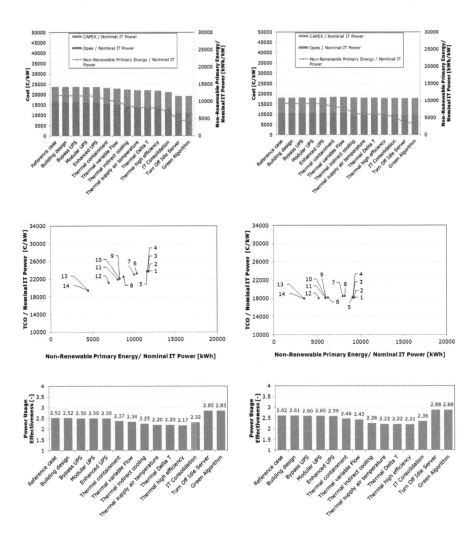

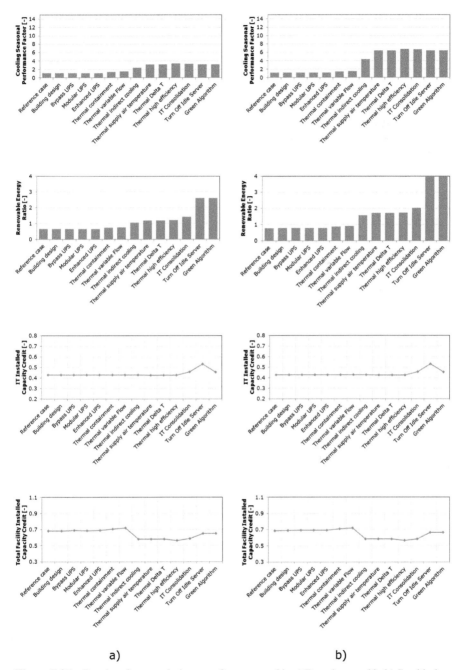

a)   b)

Figure 7.25   Results of accumulative tests for concept 6 in a) Barcelona and in b) Stockholm.

Table 7.16 Accumulation test results for the size of the main elements in concept 6

Components	Unit	Case													
		1	2	3	4	5	6	7	8	9	10	11	12	13	14
Absorption chiller (nominal power)	kW	288	288	288	288	288	288	288	288	288	288	288	288	288	288
Cooling tower (nominal power)	kW	490	490	490	490	490	490	490	490	490	490	490	490	490	490
Total air mass flow rate	t/h	768	768	768	768	768	302	302	302	302	302	302	302	302	302
Power consumption of fans	kW	94	94	94	94	94	37	37	37	37	37	37	37	37	37
Transformer (TR) (Nominal power)	kW	1870	1870	1870	1870	1870	1720	1720	1720	1720	1720	1645	1645	1645	1645
Switchgear (SWG) (Nominal current of)	kW	3569	3569	3569	3569	3569	3283	3283	3283	3283	3283	3139	3139	3139	3139
Generator (GE) (Nominal power)	kW	2500	2500	2500	2500	2500	2500	2500	2500	2500	2500	2500	2500	2500	2500
Uninterruptible Power Supply unit (UPS) (Nominal power)	kW	629	629	629	629	629	629	629	629	629	629	629	629	629	629
Power Distribution unit (PDU) (Nominal power)	kW	2500	2500	2500	2500	2500	2500	2500	2500	2500	2500	2500	2500	2500	2500

leads to decreasing values that reach nearly the reference values in both locations equally. The results of the total facility installed capacity credit show a dependency on several measures applied equally in both locations. A slight increase can be shown with the application of thermal containment, thermal variable flow, IT consolidation and turn off idle server whereas the use of thermal indirect cooling and high efficiency thermal elements leads to a decrease of the values.

When analysing the sizes of the main elements, the effect of the efficiency measures on the total air mass flow rate, the power consumption of the fans, the nominal power of the transformer and the nominal current of the switchgear show similar dependencies compared to the results for concept 1 (see Subsection 7.3.2.1).

7.3.7.2 Influence of size

The variation of the Data Centre nominal IT power shown in Figure 7.26 is evaluated for the TCO and the RER in two locations Barcelona and Stockholm. Regarding the costs it can be analysed that the decrease with the application of all energy efficiency measures is more distinctly for Barcelona (12–1%) than for Stockholm (0.2–2%). However, the absolute cost values for Barcelona are 6–25% higher. The decrease of the non-renewable primary energy values with the application of the efficiency measures reaches up to 62% equally in both locations. The dependency on the nominal IT power is comparable for both locations and lead to decreasing costs of maximal 24% for Barcelona and 25% for Stockholm while the influence on the non-renewable primary energy is negligible. The most significant decrease of the costs can be analysed from 100 kW to 500 kW (up to 14% for Barcelona and 18% for Stockholm)

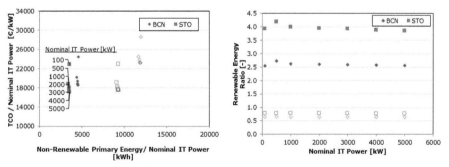

Figure 7.26 Variation of the Data Centre nominal IT power for the two locations Barcelona and Stockholm for concept 6. Unfilled shapes represent the reference case and filled shapes the case with all energy efficiency measures applied.

compared to the next variation steps (500 kW, 1000 kW, 2000 kW, etc.). Regarding the RER the application of energy efficiency measures lead to a notable increase of maximal 75% for Barcelona and 80% for Stockholm, while the size of the Data Centre just shows a minor dependency on this metric.

7.3.7.3 Influence of absorption chiller sizes and potential heat reuse

Figure 7.27 presents the results of a parametric analysis for concept 6 (Data Centre based on CHP – biogas engine driving an absorption chiller) for all the concepts. As in concept 5, the variation of two parameters has been analysed. One is the amount of heat produced by the facility which can be reused for other purposes outside the Data Centre; the other parameter is the size of the absorption chiller. The results in Figure 7.19 show the same tendency than the ones for concept 4. An increase of the absorption chiller size is not a cost-effective measure although it has an important effect on the reduction of the PE_{nren} and the increase of RER. As well as the amount of heat that can be reused is reduced (lower values of heat reuse ratio) the facility becomes more expensive to operate and PE_{nren} increases.

7.4 Other Aspects Influencing Data Centre Energy Consumption

7.4.1 Influence of the IT Load Profile

The objective of this section is to show that TCO and energy consumption are influenced by the IT load profile (how the IT facility is used). To do so, a Data Centre of 1000 kW of IT power capacity located in Barcelona is

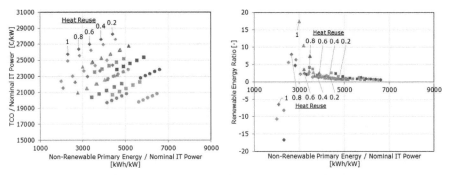

Figure 7.27 Normalized TCO (left) and RER (right) vs. normalized PE_{nren} for different scenarios, absorption chiller sizes and amount of heat reuse ratio for concept 6.

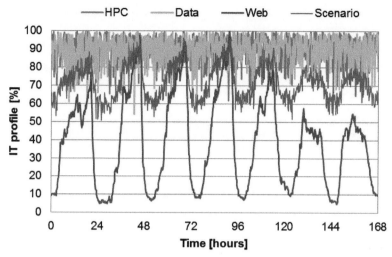

Figure 7.28 IT load profile for different scenarios.

simulated under different IT load profiles (HPC, Web, Data and mixed), as shown in Figure 7.28. Table 7.17 shows the main characteristics of the facility modelled.

Figure 7.29 shows the results (TCO and non-renewable primary energy consumption) for the four scenarios selected. Notice that in all the scenarios the CAPEX is exactly the same, so solely the IT load profile is varied. The highest energy consumption and therefore the highest TCO are experienced where HPC profile is computed, followed by the mixed scenario. This phenomenon

Table 7.17 Main parameters for the Data Centre investigated

Parameter	Unit	BCN-1000
Location	[–]	Barcelona
IT power capacity	[kW]	1000
Rack density	[kW/rack]	4
Occupancy ratio	[–]	1
Safety margin factor	[–]	0.8
White space area	[m^2]	750
IT Load profile	[–]	Mixed/Data/Web/HPC
Average electricity price	[€/kW·h$_{el}$]	0.0988
Average ratio renewables in the grid	[–]	0.36
$w_{del,total,el}$ (average)	[–]	2.29
$w_{del,nren,el}$ (average)	[kW·h$_{PE}$/kW·h$_{el}$]	1.83

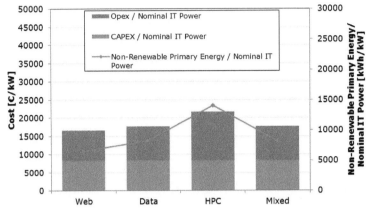

Figure 7.29 Simulation results comparing different IT load profiles.

is mainly due to the following facts: i) the IT load profile (demand) is higher for HPC ii) the power consumption associated to HPC is higher.

Figure 7.30 shows the normalized energy consumption of the IT equipment for an entire year and the normalized IT load demand. Assuming that the maximum IT load demand is where the IT load is 100% all year long, both HPC and Data load have a normalized IT load of 90% while for the purely Web load it is 40%. Notice that for the same IT load demand, a Data Centre running HPC loads consumes more energy than a Data Centre running Data load.

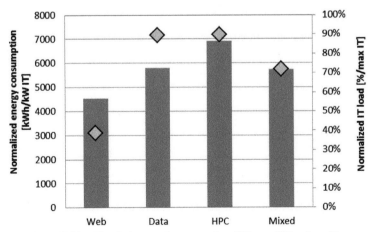

Figure 7.30 Simulation results comparing different IT load profiles.

7.4.1.1 Influence of the rack density, occupancy, and oversizing factors

This section aims to show the influence of the rack density, the occupancy ratio and the safety margin factor in the Data Centre behaviour. The TCO and the non-renewable primary energy are the metrics used to show the results and comparison between scenarios. Table 7.18 shows the main characteristics of the Data Centre analysed for a traditional energy/power supply (concept 1) without on-site renewable power systems. The safety margin factor is used to limit the maximum IT power capacity that the servers can run in the installation. As an example, a safety margin factor of 0.8 means that the maximum IT power consumption would be 80% of the total IT power capacity (1000 kW). The occupancy ratio means the ratio of installed IT, lack of occupancy is a lack of IT equipment. Therefore, for a Data Centre (power capacity of 1000 kW) with a safety margin factor of 0.8 and an occupancy ratio of 0.5, the maximum IT power consumption from the servers is 400 kW. The white space area basically depends on the nominal IT power capacity and the rack density (kW per rack). It was estimated using average ratios for floor occupied by a rack following industry used values.

Figure 7.31 shows the normalized TCO and non-renewable primary energy consumption for a constant safety margin factor of 0.8 and different values of rack density and occupancy ratio. It can be seen that when the rack density is increased the TCO is reduced due to a lower white space area needs and more compact installations. Moreover, as the occupancy ratio increases the energy consumption and therefore the TCO is increased. This phenomenon is basically because more IT load is computed in the Data Centre.

Table 7.18 Specific assumptions for the investigated Data Centre, Base case scenario for the location Barcelona varying the rack density, occupancy ratio and safety margin factor

Parameter	Unit	BCN-1000
Location	[–]	Barcelona
IT power capacity	[kW]	1000
Rack density	[kW/rack]	2 .. 10 (step 2)
Occupancy ratio	[–]	0.2 .. 1.0 (step 0.1)
Safety margin factor	[–]	0.7 .. 1.0 (step 0.1)
IT Load profile	[–]	Mixed
Average electricity price	[€/kW·h_{el}]	0.0988
Average ratio renewables in the grid	[–]	0.36
$w_{del,total,el}$ (average)	[–]	2.29
$w_{del,nren,el}$ (average)	[kW·h_{PE}/kW·h_{el}]	1.83

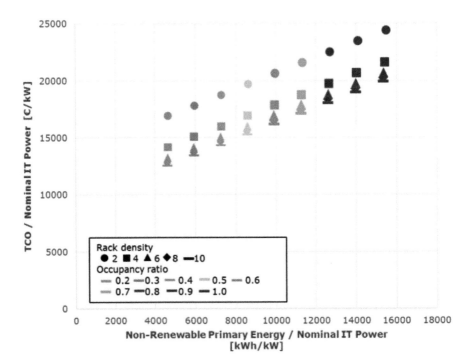

Figure 7.31 TCO and Non-renewable primary energy for different rack density and occupancy ratio using a constant safety margin factor of 0.8.

Figure 7.32 shows the normalized non-renewable primary energy in function of the safety margin factor for a constant rack density of 4 kW/rack. In this situation, the lower the safety margin factor, the lower the non-renewable primary energy consumption. This effect is due to the safety margin that limits the maximum IT power consumption, and therefore the Data Centre consumes less energy since it computes less IT load. This can also be seen in Figure 7.33, where the IT installed capacity credit is shown under the same boundary conditions. This metric shows the available IT capacity of the Data Centre, which defines the unused IT capacity and is calculated as the relation between the actual IT peak power of the system and the maximum IT peak power that the installation can handle. It is seen that for low safety margin factors, the IT installed capacity credit is higher and therefore the unused IT capacity is higher. This means that the Data Centre has a facility oversized where the PUE is higher as the safety margin factor decreases, as it is shown in Figure 7.34.

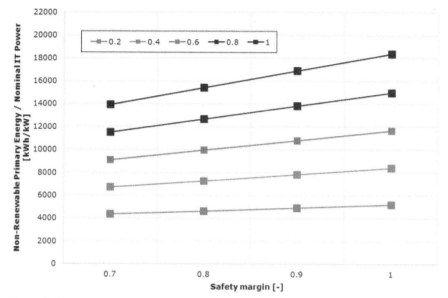

Figure 7.32 Non-renewable primary energy in function of the safety margin for a constant rack density of 4 kW/rack.

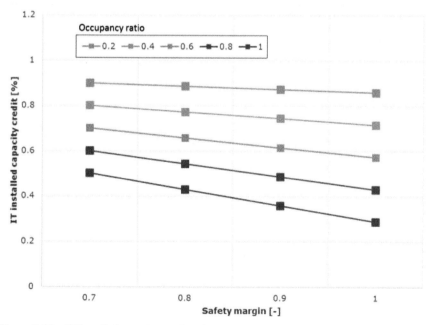

Figure 7.33 IT installed capacity credit in function of the safety margin factor and occupancy ratio for a constant rack density of 4 kW/rack.

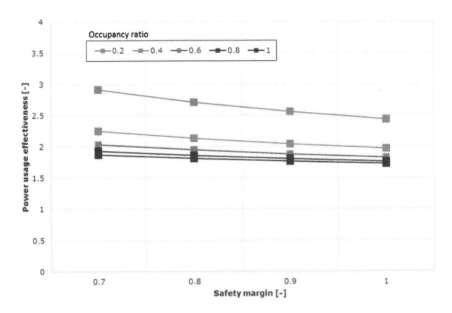

Figure 7.34 PUE in function of the safety margin factor and the occupancy ratio for a constant rack density of 4 kW/rack.

In predesign phases, instead of using the safety margin factor, the oversizing safety margin is normally used which defines the ratio between the total installed IT capacity (i.e. 1000 kW) and the maximum expected IT power consumption. The resulting extra capacity is to cover either an unexpected addition to the IT load or an unexpected impairment of the component capacity. So, the design of the power and cooling system is usually oversized in order to cover future IT load expansions. This obviously would increase the CAPEX and decrease the OPEX when working at partial loads. Figure 7.35 shows the TCO and the PUE of a facility of 1000 kW IT power capacity in function of the oversizing safety margin (from no oversized to 30% oversized) when the occupancy ratio is 80% and the average rack density is 5 kW per rack. These results clearly show how the initial investment cost (CAPEX) is higher when the oversizing safety margin is higher as well as the operational costs (OPEX) due to a reduction of the efficiency of the systems (i.e. UPS). As expected, the PUE also increases when the facility is not fitted with the IT load. These phenomenons clearly show why it is crucial to build modular Data Centres in order to reduce the TCO and increase the overall efficiency of the Data Centre.

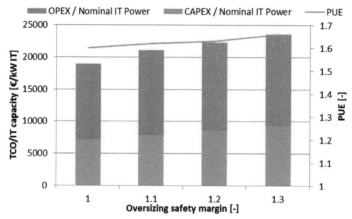

Figure 7.35 Total Cost of Ownership (CAPEX+OPEX) and PUE in function of the oversizing safety margin.

7.5 The RenewIT Tool

The RenewIT project developed a simulation tool to help Data Centre designers and operators to evaluate the energy performance of different combinations of renewable energy and efficiency measures solutions. The public RenewIT tool (www.renewit-tool.ue [6]) is a web-based planning tool to understand the costs and benefits-in economic, environmental and sustainability terms – of designing and operating a facility to use a high proportion of on-site and/or grid-based renewable energy. Therefore, the results of the tool can be used in pre-design phases and will help the industry (investors, operators, etc.) to take strategic decisions about energy efficiency strategies and renewables integration. Moreover, the RenewIT tool allows the users to analyse the benefit of implementing advanced strategies and renewables in different locations of Europe under different IT load workload and operational characteristics. The tool is built on top of advanced energy models, developed in TRNSYS. However, the RenewIT tool, as any other web application, needs to process the information as fast as possible in order to enhance the user experience. Therefore, it has been necessary to generate metamodels which are the computational core of the tool. Metamodels are simplified models generated by surface-response methods from the detailed dynamic models which runs with only key parameters already identified.

The Graphical User Interface (GUI) is divided in six different sections: general information, IT infrastructure, power and cooling characterization,

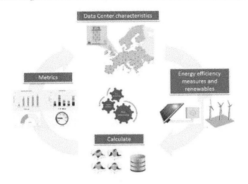

Figure 7.36   Home page of the RenewIT tool in www.renewit-tool.eu.

selected options and results page. Figure 7.37 shows an image of the different sections and the most relevant information requested in each stage. Behind the GUI there is the engine which processes the information introduced by the user combined with data from data bases and calculates the results for each scenario using the metamodels library.

7.6 Conclusion

Data Centres are very complex energy facilities which are influenced on one hand for the characteristics of the Data Centre such as type of IT load, cooling system, power distribution and on the other hand for boundary conditions such as location (weather and local energy resources and energy infrastructures availability), electricity price, etc. Therefore, dynamic energy modelling is necessary to predict energy and economic behaviour of these unique infrastructures.

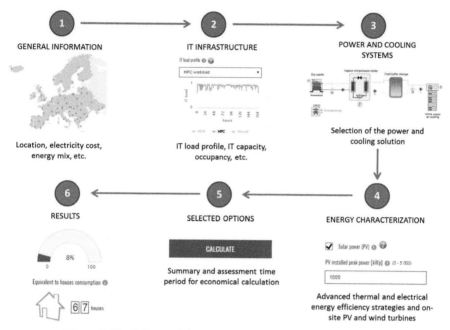

Figure 7.37 Scheme of the RenewIT tool functionality and layout.

This chapter provided a basic parametric analysis for each of the 6 concepts described varying some of the main parameters of each advanced solution. Concepts based on biogas CHP systems (with fuel cells or reciprocating engine) present promising numbers of PE_{nren} reduction but are not cost effective. The connection of a Data Centre to a district heating and cooling network while reusing the heat from liquid cooled servers is a very promising solution and cost effective. Using on-site PV or wind power production can be an interesting option towards Net Zero Energy Data Centres, but is dependent on the location.

The results shown and described in this chapter give enough information to know the influence of the variation of the main parameters of influence. However, it is needed to use detailed optimization and dynamic energy models together with information of local constraints, as energy prices or available space for renewables, to seek for a cost effective option and/or environmentally friendly solution.

References

[1] TRNSYS Thermal Energy System Specialists LLC, Version 17. [Online]. Available: http://www.trnsys.com/

[2] Wikipedia, "Köppen climate classification," 2015.

[3] CEN, "EN 15459 Energy performance of buildings – Economic evaluation procedure for energy systems in buildings," 2007.

[4] A. O. E. S. J. C. M. M. M. G. J. Carbó, "Experimental and numerical analysis for potential heat reuse in liquid cooled data centres," Energy Conversion and Management, vol. 112, pp. 135–145, 2016.

[5] Meteotest, "Meteonorm software v7," [Online]. Available: www.meteonorm.com. [Accessed 2017].

[6] "RenewIT tool," 2016. [Online]. Available: http://www.renewit-tool.eu/.

[7] IDAE, "CO2 emission factors," 2011.

[8] P. d. l'Alba. [Online]. Available: http://www.parcdelalba.cat. [Accessed 2017].

Annexes

Annex 1. Nomenclature

Energy Flows Nomenclature

This section summarizes the nomenclature for the main energy flows used in the examples (Section 3.6).

W_{GRID}	Electrical energy from grid.	Represents the electrical energy purchased from the electrical grid.
W_{PV}	Electrical energy generated by the on-site photovoltaic plant.	Represents the electrical energy delivered to the data centre by the on-site photovoltaic plant.
W_{CHP}	Electrical energy generated by the on-site combined heat and power plant.	Represents the electrical energy delivered to the data centre by the on-site combined heat and power plant.
$W_{E,SUB}$	Electrical energy in the electrical substation.	Represents the electrical energy which enters to the electrical substation.
$\zeta_{E,SUB}$	Electrical losses in the electrical substation.	Represents the electrical losses due to the conversions and inefficiencies of the systems contained in the electrical substation (transformers, switch gears, etc.)
$W_{E,PDTS}$	Electrical energy in the power distribution system.	Represents the electrical energy which enters to the data centre facility after the conversions and losses in the electrical substation.
$W_{E,MISC}$	Electrical energy consumption of the miscellaneous systems.	Represents the electrical energy consumed by the miscellaneous systems in the data centre (lighting, appliances, etc.)
$W_{E,MTS}$	Electrical energy consumption of the mechanical technical systems.	Represents the electrical energy consumed by the cooling equipment (VCCH, ABCH etc.) of the data centre and as well the electrical consumption of the auxiliary cooling systems such as fans, pumps and controls. If the system uses free cooling strategy the energy consumption for free cooling must be also considered.

(Continued)

241

Annex 1. Continued

$\zeta_{E,PDTS}$	Electrical losses in the power distribution systems.	Represents the electrical losses due to the conversions and inefficiencies of the power distribution systems contained in the data centre.
$W_{E,IT}$	Electrical energy consumption of the IT equipment.	Represents the electrical consumption of the IT equipment in the data centre.
$Q_{C,chw,1}$	Cooling energy delivered by the cooling equipment.	Represents the heating removed due to the cold produced by the cooling equipment of the data centre such as VCCH, ABCH, HP etc.
$Q_{C,chw,2}$	Cooling energy delivered by the district cooling network.	Represents the heating removed due to the cold purchased in the district cooling network.
$\xi_{th,MTS}$	Thermal losses in the mechanical technical systems.	Represents the thermal losses due to the thermal inefficiencies introduced by conversion and transport of cooling (such as losses in CRAC, CRAH, pipes, extra pumps or fans) and inefficiencies due to air management in IT rooms
$Q_{C,FC}$	Cooling energy delivered by the free cooling strategy.	Represents the heating removed in the IT room with the outside air when the free cooling strategy is applied.
$Q_{C,IST}$	Cooling energy stored in the ice storage tank.	Represents the cooling stored, for future use in the data centre, by the ice storage tank.
$Q_{C,IT}$	Cooling energy delivered in the IT room.	Represents the final heating removed in the IT room due to the cooling equipment.
$Q_{H,ABCH}$	Heating energy delivered to the cooling equipment.	Represents the heating delivered to produce cold with an absorption compression chiller.
$Q_{H,DH}$	Heating energy exported or imported to a district heating network.	Represents the heating exported or imported by the data centre to the district heating network.

Annex A. Subsystems for Advanced Technical Concepts of Cooling and Power Supply

A.1 Introduction

In the present annex, descriptions of the subsystems for advanced technical concepts of cooling and power supply are provided. The subsystems are divided into the following categories: power generation, cooling production, energy storage, heat production and waste heat recovery.

Detailed information on the various technologies can be found in the deliverable 4.1 of the project RenewIT (*Report of different options for renewable energy supply in Data Centres in Europe* [42]).

A.2 Power Generation

A.2.1 Photovoltaics

General Description

The key components of a photovoltaic (PV) power system are various types of PV cells (also known as solar cells), which are interconnected and encapsulated to form a PV module (the commercial product), the mounting structure for the module or array, the inverter (essential for grid-connected systems and required for most off-grid systems), the storage battery and charge controller (for off-grid systems but also increasingly for grid connected ones).

The PV modules have to be integrated through structures that can be mounted directly onto roofs or integrated in the buildings (building integrated PV, BIPV), including PV facades, integrated (opaque or semi-transparent) glass–glass modules and "PV roof tiles". Tracking systems have recently become more and more attractive, particularly for PV utilisation in countries with a high share of direct irradiation. By using such systems, the energy yield can typically be increased by 25–35% for single-axis trackers and by 35–45% for double-axis trackers compared with fixed systems. However, falling costs in PV cells and maintenance criteria might not make them the best option.

Scheme

The scheme of grid-connected PV system and overvoltage protection is shown in Figure A.1.

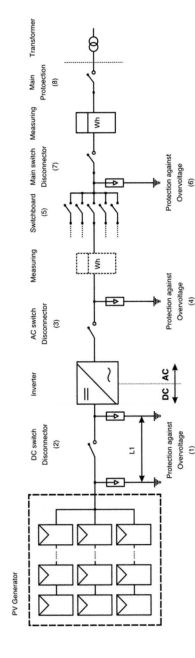

Figure A.1   Example of grid-connected PV system and over-voltage protections. PV solar system connected to the local switchboard enables self-consumption of PV solar electricity.

Control

A PV system can be connected to the grid (grid tie) or be independent (off-grid). In both cases, a control system is required to regulate and convert the generated power. The main element responsible for this task is known as inverter. Grid-connected inverters must supply electricity in sinusoidal form AC, synchronised to the grid frequency, limit feed-in voltage to not be higher than the grid voltage and disconnect from the grid if the grid voltage is turned off. Islanding inverters only need to produce regulated voltages and frequencies in a sinusoidal wave-shape as no co-ordination or synchronisation with grid supplies is required.

Calculation

The operation of PV systems depends on the solar irradiation of the site and the technical conditions of the system. Below, a simple methodology for the calculation of grid-connected PV systems is described.

The nominal power of a photovoltaic system is the sum of the nominal power of the modules. The nominal power is the power delivered by the modules at AM1.5 (1000 W/m^2) radiation and 25°C of module operation temperature.

If a module is irradiated by H kWh/m^2, the energy of the radiation received by the module is equivalent to H hours of normal AM1.5 radiation. Therefore, a first simple estimation of the production can be made by the following equation:

$$E = P_{nom} \cdot H$$

The equivalent number of hours of normal radiation is available for a large number of locations [28].

The area required to achieve the desired nominal power depends on the efficiency of the PV module in the range of 110 to 145 W_p/m^2.

However, the performance of a system under real operation conditions differs from the nominal efficiency of the modules. This performance ratio (*PR*) of the system can be separated in two factors: the performance ratio of the module (PR_{mod}) and the performance ratio of the balance of system (PR_{BOS}). With the performance ratios, the electric energy (E) produced by the system can be calculated by the following equation:

$$E = P_{nom} \cdot PR \cdot H = P_{nom} \cdot PR_{mod} \cdot PR_{BOS} \cdot H$$

Due to sun radiation, the modules operate at a higher temperature than ambient temperature and it is necessary to introduce a temperature correction of PR.

The variation of the module power due to an operating temperature different from standard test conditions temperature (25°C) can be determined by using the temperature coefficient of the module power (α_{PT}):

$$PR_{\mathrm{mod}} = 1 + \alpha_{\mathrm{PT}} \cdot (T_{\mathrm{amb}} - 25 + T_{\mathrm{NOCT}} - 20)$$

The rest of the system components (BOS) also introduce losses. Cabling is responsible for resistive energy losses that usually are kept lower than 1% during the design phase. The modern inverters are very efficient and present a small amount of losses in the range of 2% to 6%. If there are transformers to change voltages, it is also usual to have 2.5% losses for every transformer step. Differences in the performance of the modules produce a mismatch loss that is usually in a 1 to 3% range. In a linear approximation, all these effects can be introduced as a constant PR_{BOS} usually in the 0.85 to 0.90 range.

Limits of Application

PV systems do not have any specific limitation, except for the space required for its installation. However, this technology makes sense only in those locations where the radiation levels justify the investment.

Economic Aspects

The cost in the case of PV largely depends on the size of the plant to be installed. The information regarding the differences in costs between the type of system installed and the cost of the different components in the global system cost is given in [18]. It can be noticed that the percentage of cost of the module in the global cost oscillates between a 30% and a 40% of the global cost (0.8–2.3 €/Wp), and the trend in the last years is a continuous reduction in costs. In the case of Data Centres, when talking about either commercial rooftop or utility mounted on the ground (either fixed tilt or one axis tilt), costs can oscillate between 2.3 and 3.4 €/Wp.

A.2.2 Wind Turbines

General Description

Wind energy is the conversion of the kinetic energy of wind into a useful form of energy, like mechanical energy (pumping) or electricity (through the use of wind turbines integrated with electric generators). Wind turbines are the main element harvesting this kinetic energy. These can be classified into the following two groups:

- Horizontal axis machine (HWTs): Rotation axis is parallel to wind direction, in a similar way as the classic windmills.
- Vertical axis machine (VWTs): Rotation axis is perpendicular to wind direction. The most common types are the Savonius and Darrieus wind turbine.

Actually, horizontal axis generators are most common as electricity generators in big wind farms.

On the other hand, wind turbines placed in urban locations are characterised by a lower annual mean wind speed (AMWS) compared with rural areas, and more turbulent flow [2]. The lower velocity is caused by the obstructions around the turbine that do not allow the wind to increase its velocity. Wind flow becomes turbulent because of these obstructions.

In order to supply energy to a Data Centre, it is important to consider the location of the energy production plant. It is clear that HWTs are more efficient than VWTs in clear spaces with laminar flow. However, lower wind speeds, turbulent flow and variable wind direction (city conditions) can favour the use of VWTs.

Control

Flexibility of operation is conditioned to the wind resource at any moment. Wind turbines have modern control systems that allow a wide wind speed range with a similar power ratio, so there are fewer turbine limitations in flexibility terms. Anyway, all the wind turbines present a cut in and cut out speed to work, so flexibility is compromised by wind sources.

Control of the rotational speed of the turbine is done through resistors, disorientation or changing the blade angle to avoid accidents if the wind exceeds recommended limits. Modern turbines with blades are usually pitch active controlled. That means that blades change their pass angle to reduce wind harvesting. There is also another control type that is based in the aero dynamical design of the blades, known as stall control.

Calculation

Wind energy depends basically on the capacity of the turbine for harvesting the wind power, the maximum amount of power being the amount of kinetic energy of the wind, which depends on the area covered by the turbine blades

(A_R), the density of the air (ρ), the velocity of the wind (V) and the power coefficient (C_p):

$$P = \frac{1}{2}C_P\rho A_R V^3$$

However, not all the kinetic energy of the wind can be harvested. The maximum physical efficiency of the system according to Betz's law is 59% of the total kinetic energy of the wind.

High-speed turbines with horizontal axis have a maximum C_p value about 0.4 at wind speeds 7 or 10 times greater than for low-speed turbines. A VWT Savonius rotor has a maximum C_p value about 0.3 at about 1 m/s wind speed. Darrieus rotor has a maximum C_p value greater than Savonius at about 4–6 m/s wind speed.

In addition to this power coefficient, losses due to mechanical friction and electronic converter must be added. The gearbox performance (η_t) is over the 80%, and the electric generator (η_m) is over the 95%. This means that final electric conversion efficiency is between 25 and 33% in most cases.

In order to know the wind speed on a location, it is necessary to have a probability distribution. A variety of probability density functions have been used in literature to estimate wind energy potential, but the Weibull function is most widely adopted.

Using the above expression and the values of Figures A.2 and A.3, it is possible to calculate the power generated during a year.

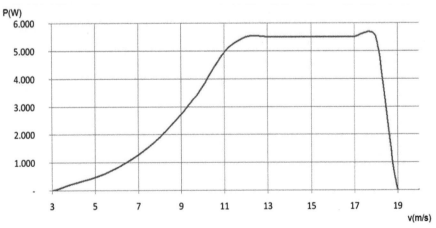

Figure A.2   Performance of a micro wind turbine Tornado (5 kW) [48].

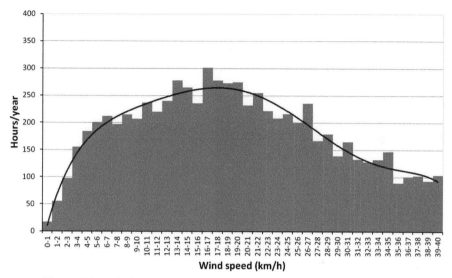

Figure A.3 Wind speed distribution in a station of Montjuïc (Barcelona) [1].

Limits of Application

The location of the wind turbine has a big influence on the power range. The mean ratio power of the urban HWT is about 9 kW with a rated wind speed about 12 m/s. For the urban VWT, the mean ratio power is a bit lower, about 7 kW with a higher rated wind speed of 13 m/s.

The mean ratio power of open source HWT is about 2.3–3.0 MW with a mean rated wind speed of 12 m/s. Mean cut in speed is about 3 m/s, and cut out speed is about 24 m/s.

Offshore HWT mean power ratio is from 3.0 to 8.0 MW. The mean rated wind speed is similar than HWT on shore, about 12–13 m/s. The same for the cut in and cut out speed, being the main differences not in operation but in structural placement and costs.

Economic Aspects

The cost of wind power plants depends on the type of placement (onshore/offshore) [33]. For onshore technology, estimated investment cost is about 1,450 €/kW + 25% [33]. Cost improvements of this maturing technology through 2050 are not expected. Performance improvements due to new technologies in the wind turbine are expected until 2030; further improvements are not assumed to be achievable after 2030 (Table A.1).

Table A.1   Summary of available wind turbine power ranges

Location	Name	Power Range (kW)
Onshore	Micro turbine (very low power)	0–3
	Mini turbine (low power)	3–50
	Meso turbine (medium to high power)	50–1000
	Macro turbine (very high power)	1000–3000
Offshore	Meso turbine (medium to very high power)	50–1000
	Macro turbine (very high power)	1000–8000
Urban	Urban turbines vertical axis	0.025–100
	Urban turbines horizontal axis	0.4–30

For fixed bottom offshore technology, estimations of investment costs are 2,400 €/kW + 35% [33]. Until 2030, performance and 10% of cost improvement are assumed to be achievable.

The offshore wind technology with floating platforms are suitable in deep water because a tower and foundation are not cost effective in this case. This technology is under development right now and allows high power production. In 2020, investment cost is expected to be 3,000 €/kW + 35% [33]. Cost and efficiency improvements of 10% are expected through 2030.

Urban wind turbines are not as common as the ones described before, so it is more difficult to collect cost data. It is assumed that urban wind turbine costs ranges from 6,000 €/kW to 10,000 €/kW [8]. Costs can be very variable depending on the connection configuration of the urban wind turbines, if their installation is part of a stand-alone system, or it is necessary to pay for a grid connection.

A.3 Power and Heat Production

A.3.1 Fuel Cells

General Description

Fuel cells (FC) are electrochemical devices that convert the chemical energy of fuel, such as natural gas or hydrogen, into direct current and heat energy without any combustion (but oxidation) process. The produced electricity can be used to power Data Centres, and heat can be used to produce steam for CHP plant or for cooling purposes through absorptions chillers.

The working temperature of fuel cell varies from $-10°C$ [24] to $1000°C$ [25], and the total efficiency can reach up to 90% if heat is used, depending on the application [19]. The basic elements of fuel cells comprise of anode, which requires H_2 fuel, cathode, which brings O_2, and electrolyte, where exchange of ions takes place [22, 23, 29].

There are different fuel cell typologies depending on its electrolyte. The most common types include [43] proton exchange membrane (PEM), phosphoric acid fuel cell (PAFC), direct methanol fuel cell (DMFC), alkaline fuel cell (AFC), direct carbon fuel cell, solid oxide fuel cell (SOFC) and molten carbonate fuel cell (MCFC).

Some companies, such as Bloom Energy, Hydrogenics Corp, and United Technologies and Fuel Cell Systems (UPS Systems), produce fuel cells for Data Centres [4]. There are several ways to place fuel cells in the Data Centre. For example, that can be at the utility power level, at the rack level and at the server level [43]. However, their energy source (natural gas) is generated off-site and distributed by a utility [4]. For the smooth operation of fuel cell, the reliable supply of fuel is utmost important and additional equipments such as reformers, battery systems, start-up systems and auxiliary circuits are needed [43]. Figure A.4 illustrates different processes in obtaining hydrogen from different carbon-based fuels.

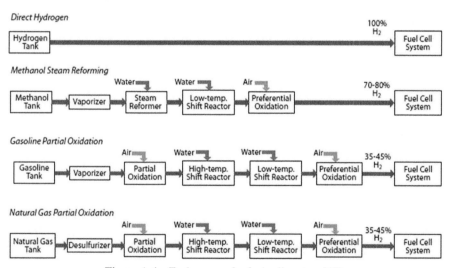

Figure A.4   Fuel sources for fuel cell system [47].

Hydraulic Scheme

Figure A.5 represents the hydraulic scheme of a fuel cell system.

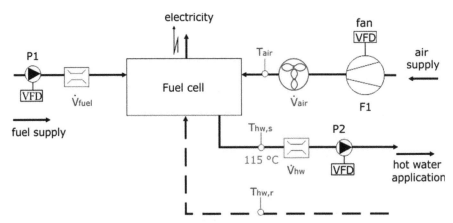

Figure A.5   Hydraulic scheme of a fuel cell (example temperature).

Calculation

Given below are some of the important performance parameters of fuel cells.

The energy conversion efficiency of the fuel cell can be calculated using the following equation [46]:

$$\eta_{FC} = \frac{P_{FC}}{\dot{m}_{fuel}\ LHV}$$

Here, P_{FC} is the electrical power output, $\dot{m}_{fuel}$ is the flow rate, and *LHV* is the lower heating value of H_2.

Capacity factor of a fuel cell shows the ratio of the electricity generated, for a certain period, to the energy generated at full power operation during the same period [29]. Thermal efficiency of a fuel cell shows the ratio of thermal heat available to the total amount of energy put in.

Table A.2 lists characteristics of two promising types of fuel cells, SOFC and PEM.

Control

A satisfactory transient behaviour of fuel cell is very crucial for its smooth operation in Data Centre. Four main subsystems of fuel cell plants are reactant

Table A.2 SOFC and PEM characteristics [34, 35, 43]

	SOFC	PEM
Size range	2 kW–200 MW	2 kW–200 kW
Catalyst	Peroktives	Platinum
Electrolyte	Ceramic	Ion exchange membrane
Fuel	Natural gas, H_2, coal-derived gas	Pure H_2
Electric efficiency	50%–60%	40%
Operating Temperature	800–1000°C	80–150°C
Typical application	High-power generation, stationary, transport	Low-power generation, space, transport, stationary
Advantages	high efficiency, CHP	rapid start-up, fast on/off

supply; heat, temperature, pressure and humidity control; water management; and power management [43]. A control system manages correctly these subsystems in stack [43].

Part Load Characteristics

Figure A.6 shows the variation of overall energy efficiency under different load conditions. At little load, the efficiency is too low, and as load increases,

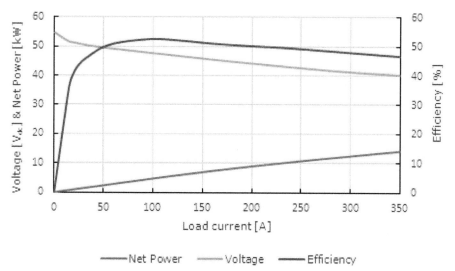

Figure A.6 Voltage, net power and efficiency of a 10 kW PEM under different load conditions [43].

efficiency decreases marginally. Therefore, FCs operate insufficiently at too little and too much load [43]. The power consumption of Data Centres fluctuates with the variation of workload and server on/off [43]. SOFC has very slow response time (e.g., 200 W/min) [5]. Therefore, to cope with this variation, FCs need to be over-provisioned or powered by external batteries [43].

Limits of Application

The main limitations of this technology are the fuel availability and the space required to store it.

Economic Aspects

The cost of FCs at present is high due to the limited production, around 25,000 units/year globally [43]. Reduction of cost is expected due to the invention of cheaper material and mass production. The conservative cost range of fuel cell is 3–5 \$/W (2.2–3.6 €/W)[1] [43].

A.3.2 Reciprocating Engine CHP

General Description

Reciprocating internal combustion engines are a widespread and well-known technology. There are two basic types of reciprocating engines for CHP uses: spark ignition (otto cycle) and compression ignition (diesel cycle). Both use a cylindrical combustion chamber with a piston. After ignition, the piston is pushed by the gases and moves the crankshaft, converting the linear movement into a circular one. The main difference between the two designs is that otto cycle needs to ignite the mixed fuel with a spark, and the diesel ones make ignition due to high pressure and temperatures of the chamber.

There are four sources of usable waste heat from a reciprocating engine: exhaust gas, engine jacket cooling water, lube oil cooling water and turbocharger cooling. Recovered heat generally is in the form of hot water. The high-temperature exhaust can generate medium pressure steam, but the hot-exhaust gas contains only about one half of the available thermal energy from a reciprocating engine. Some industrial CHP applications use the engine exhaust gas directly for process drying. Generally, the hot water produced by reciprocating engine CHP systems (Figure A.7) is appropriate

---

[1]1\$ = 0.73 € (Exchange rate as on 26 March 2014).

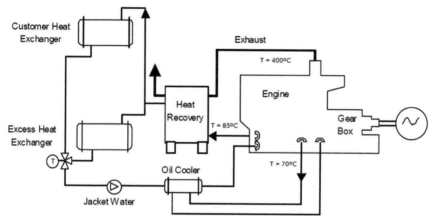

Figure A.7 Closed-loop heat recovery system in a reciprocating engine (example temperatures) [51].

for low-temperature process needs, space heating and potable water heating, and to drive absorption chillers providing chilled water, air conditioning or refrigeration.

Control

Reciprocating engines are controlled by the amount of fuel injected in the CHP system. The control depends on which is the priority demand to supply. However, thermal and electrical demand can be supplied simultaneously. If thermal energy is needed during CHP operation, there is no problem. Heat demand is supplied while electricity can be delivered to the grid or can be consumed by the process or the building. If thermal demand is not needed during CHP operation, there must be a heat storage device. Heat transportation is expensive and has a reduced efficiency due to the thermal losses. Other way, this heat storage device has a limited capacity. If there is no heat demand and the heat storage device is already full, CHP system may stop its operation in order not to waste heat. Mostly, an emergency or auxiliary cooler is installed for operational safety.

Calculation

The total efficiency of the plant is defined as follows:

$$\eta_{\text{CHP}} = \frac{\dot{Q}_H + P_{\text{el}}}{\dot{m}_{\text{fuel}} \, \text{LHV}}.$$

where P_{el} is the electric power, $\dot{Q}_H$ is the useful rate of heat, $\dot{m}_{fuel}$ is the rate of supplied fuel, and *LHV* is the lower heating value of the fuel. The total efficiency is often divided into the two parts, instantaneous electric efficiency $\eta_{el,CHP}$ and thermal efficiency $\eta_{th,CHP}$ with $\eta_{CHP} = \eta_{el,CHP} + \eta_{th,CHP}$:

$$\eta_{el,CHP} = \frac{P_{el}}{\dot{m}_{fuel}\,LHV}; \quad \eta_{th,CHP} = \frac{\dot{Q}_H}{\dot{m}_{fuel}\,LHV}.$$

Limits of Application

The CHP plant requires a certain amount of annual operating hours in order to make economic sense. Thus, there must be an appropriate heat demand available close to the Data Centre, which absorbs the rejected heat especially during winter, when the Data Centre is cooled by means of indirect air free cooling.

Economic Aspects

The investment costs (without gas processing) of a reciprocating engine for biogas CHP can be estimated by [6]

$$c_{\dot{Q}} = f_{RPI}^{(y-2011)}\left(15648\,P_{el}^{-0,536}\right)$$

where $c_{\dot{Q}}$ und P_{el} represent the specific investment costs in €/kW and the electric power in kW, respectively. The factor f_{RPI} indicates the mean annual rate of cost increases (e.g., 1.03) from 2011 to the current year y.

A.4 Cooling Production

A.4.1 Absorption Chiller

General Description

An absorption chiller produces chilled water by transferring heat from the chilled water circuit to the re-cooling circuit (closed cycle). The absorption cycle is driven by heat supplied to the generator.

The main components of an absorption chiller are evaporator, condenser, generator and absorber as well as expansion devices and a solvent pump as shown in Figure A.8. For HVAC applications and chilled water production, water and lithium bromide are most commonly used as refrigerant and solvent, respectively [53]. Absorption chillers are available as single-effect,

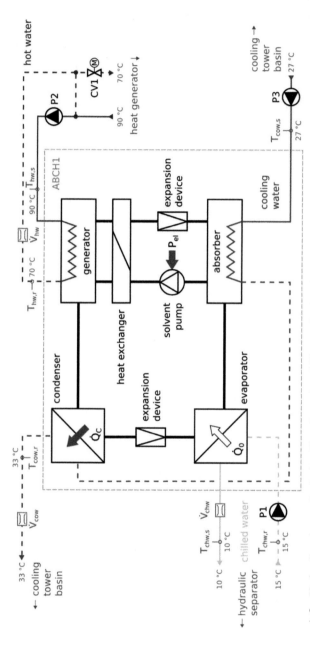

Figure A.8 Hydraulic scheme of a single-effect absorption chiller (H_2O-LiBr) for water cooling; re-cooled with water (example temperatures; scheme based on [53]).

double-effect or triple-effect machines. Double- or triple-effect chillers can be applied when high driving temperatures are available and lead to higher *COP*s compared to single-effect absorption chillers.

In addition to the chiller itself, the three pumps P1, P2 and P3 shown in Figure A.8 are considered as parts of the subsystem.

Hydraulic Scheme

The hydraulic scheme of a single-effect absorption chiller is shown in Figure A.8.

Control

Cooling power of the chiller is controlled internally for reaching the set point of chilled water supply temperature $T_{chw,s}$. Constant flow through the shell-and-tube heat exchangers is recommended. Thus, it is proposed to run an absorption chiller with constant cooling power and use a thermal energy storage for levelling fluctuating loads.

In the hot water circuit, a variable supply temperature $T_{hw,s}$ can be reached by controlling the addition of return water to the supply by means of a control valve CV1.

Calculation

The heating power $\dot{Q}_H$ needed to produce a given cooling power $\dot{Q}_0$ can be calculated from the coefficient of performance (*COP*):

$$\dot{Q}_H = \frac{\dot{Q}_0}{COP}$$

The *CoP* depends, for example, on the construction type of the chiller, on the current usage rate and on the media used as refrigerant and solvent. Typical *COP* values are shown in Table A.3. Detailed rating values are available from literature, e.g. [13].

Absorption chillers are characterised by a beneficial part load behaviour. Normally, cooling power can be reduced to 10% of the rated power as shown in Figure A.9.

As it is shown by the energy flow scheme in Figure A.10, the heat flow $\dot{Q}_{con} + \dot{Q}_{Abs}$ which is to be removed by the re-cooling circuit amounts to

$$\dot{Q}_{con} + \dot{Q}_{Abs} = \dot{Q}_0 + \dot{Q}_H.$$

Table A.3 Characteristics data of absorption chillers for the production of chilled water and process cooling [53]

	H$_2$O–LiBr ABCH (SE)	H$_2$O–LiBr ABCH (DE)	NH$_3$–H$_2$O ABCH
Application	Chilled Water	Chilled Water	Process Cooling
Chilled water outlet temperature [°C]	5–25	5–25	–50–5
Cooling water inlet temperature [°C]	16–45	16–45	
Inlet temperature of heating medium [°C]	75–140	140–170	100–160
COP	0.55–0.75	0.8–1.2	0.35–0.65

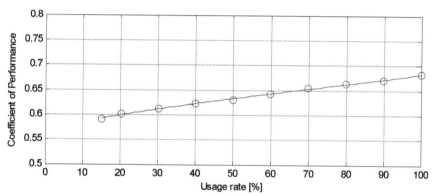

Figure A.9 Typical part load behaviour of LiBr-H$_2$O absorption chillers (single-effect) [21].

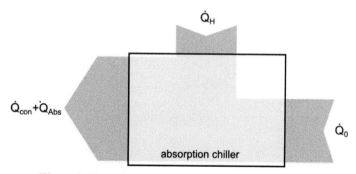

Figure A.10 Scheme of heat flow in an absorption chiller.

The specific internal auxiliary energy demand for pumps accounts for $E_{aux} = 15\ \mathrm{Wh/kWh_C}$ [13].

Pumping hot water, chilled water and cooling water through the chiller requires electrical energy. The power P_{Pump} required for a given volume flow rate $\dot{V}$ can be calculated using the following equation:

$$P_{pump} = \frac{\dot{V}_{chw} \cdot \Delta p_{chw}}{\eta_{pump}} \tag{A.1}$$

Limits of Application

Absorption chillers cover a wide range of cooling power from some kW to some MW. Their application is mainly limited by the availability of a suitable heat sink for re-cooling. For driving the chiller, cheap heat must be available with a temperature of at least 75°C or 140°C (single-effect or double-effect H_2O-LiBr chiller, respectively [53]). Generally, absorption chillers should be used when at least 90°C are available while adsorption chillers are recommended for lower temperatures. The cooling water supply temperature must not drop below 25°C [45].

Economic Aspects

The investment costs of a single-effect absorption refrigerator (without re-cooling plant) can be estimated by [20]

$$c_{\dot{Q}} = f_{RPI}^{(y-2002)} \left(14{,}740.2095\ \dot{Q}_0^{-0.6849} + 3.29 \right)$$

$$\left(\text{valid for } 50\ \mathrm{kW} \leq \dot{Q}_0 \leq 4{,}750\ \mathrm{kW} \right)$$

where $c_{\dot{Q}}$ and $\dot{Q}_0$ represent the specific investment costs in €/kW and the cooling power in kW, respectively. The factor f_{RPI} indicates the mean annual rate of cost increases (e.g., 1.03) from 2002 to the current year y.

For a double-effect machine, the cost function reads [20]

$$c_{\dot{Q}} = f_{RPI}^{(y-2002)} \left(231975.0507\ \dot{Q}_0^{-1,1422} + 90.09 \right)$$

$$\left(\text{valid for } 400\ \mathrm{kW} \leq \dot{Q}_0 \leq 5270\ \mathrm{kW} \right).$$

A.4.2 Vapour-Compression Chiller

General Description

A vapour-compression chiller produces chilled water by transferring heat from the chilled water circuit to the re-cooling circuit. The vapour-compression cycle is driven by electrical energy supplied to the compressor. Re-cooling can be realised either by water-cooling or by air-cooling.

The main components of a vapour-compression chiller are evaporator, compressor, condenser and expansion device as shown in Figure A.11. Many different construction types are available; an overview is given for example in [14].

In addition to the chiller itself, pumps are needed for the chilled water and the cooling water circuit.

Hydraulic Scheme

The hydraulic scheme of a vapour-compression chiller is shown in Figure A.11.

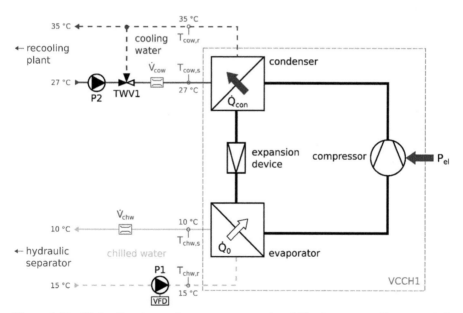

Figure A.11 Hydraulic scheme of a vapour-compression chiller for water cooling; re-cooled with water (example temperatures; scheme based on [53]).

Control

The cooling power is controlled internally in order to reach the set point of the chilled water supply temperature $T_{chw,s}$. A variable speed pump P1 can be used for adjusting the chilled water flow to the cooling load. With state-of-the-art chillers, manufacturers allow variable evaporator flow as long as minimum and maximum flow velocities as well as a maximum rate of chilled water flow variation are met [7]. If the evaporator flow rate becomes too small, heat transfer may drop severely.

For achieving a constant cooling water supply temperature $T_{cow,s}$ which is not too low, a three-way valve TWV1 can be used for admixing return water to the supply.

Calculation

The electrical power P_{el} needed to produce a given cooling power $\dot{Q}_0$ can be calculated from the coefficient of performance:

$$P_{el} = \frac{\dot{Q}_0}{COP}.$$

The coefficient of performance depends, for example, on the construction type of the chiller and on the current usage rate and generally has a value of 3 to 8 in HVAC applications [20]. Detailed rating values for different compressor types, refrigerants and so on are available from literature, e.g., [13].

Cooling power of vapour-compression chillers can be reduced to 20% of the rated power or less, but *COP* decreases as well in this case as shown in Figure A.12.

As it is shown by the energy flow scheme in Figure A.13, the heat flow $\dot{Q}_{con}$, which is to be removed by the re-cooling circuit, amounts to

$$\dot{Q}_{con} = \dot{Q}_0 + P_{el}.$$

As far as the pump energy demand is concerned, please refer to Section A.4.1.

Limits of Application

Vapour-compression chillers cover a wide range of cooling power from some kW to some MW. Their application is mainly limited by the availability of a suitable heat sink for re-cooling. The cooling water supply temperature must not be too low. The minimum value depends on the process and the machine as well as the refrigerant.

Economic Aspects

The investment costs of a vapour-compression chiller (without re-cooling plant) can be estimated by [20]

$$c_{\dot{Q}} = f_{\text{RPI}}^{(y-2002)} \left(4{,}732.2487 \, \dot{Q}_0^{-0.7382} + 109.30 \right)$$

$$\left(\text{valid for } 10 \text{ kW} \leq \dot{Q}_0 \leq 10{,}000 \text{ kW} \right)$$

where $c_{\dot{Q}}$ and $\dot{Q}_0$ represent the specific investment costs in €/kW and the cooling power in kW, respectively. The factor f_{RPI} indicates the mean annual rate of cost increases (e.g., 1.03) from 2002 to the current year y.

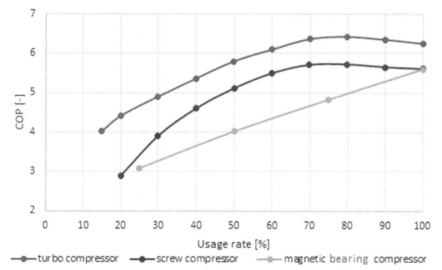

Figure A.12 Typical part load behaviour of vapour-compression chillers depending on the compressor type [11, 21].

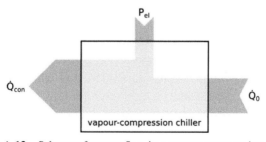

Figure A.13 Scheme of energy flow in a vapour-compression chiller.

A.4.3 Evaporative Cooling

General Description

Evaporative cooling process (adiabatic cooling) is commonly used in cooling water towers, air washes, evaporative condensers and fluid cooling and to soothe the temperature in places where several heat sources are present. It is a simple and effective way of cooling an air stream.

Evaporative cooling equipment can be direct evaporative cooler or indirect cooler. In a direct evaporative cooler (DIEC), the air stream to be cooled is in direct contact with a liquid water film and cooling is accomplished by the adiabatic heat exchange between the air stream and the liquid water film. The evaporation of water in the air stream leads to a reduction in the dry-bulb temperature. However, this will also concurrently cause an increase in the humidity ratio of the air stream. Generally, the maximum possible reduction in air dry-bulb temperature depends on the difference between the dry-bulb and the wet-bulb temperature of the air stream. In many applications, like Data Centres, the increase in humidity in the supply air stream is not desirable. In this case, an indirect evaporative cooling system can be employed.

In an indirect evaporative cooling system, the process air (primary airflow) transfers heat to a secondary airflow or to a liquid that has been cooled by evaporation. By using a cross-fluted heat exchanger, the water never comes in contact with the air. In this case, both dry-bulb and wet-bulb temperatures are reduced.

Using evaporative cooling (DIEC) for the first stage of cooling make-up air reduces energy costs. The second stage is handled by conventional air conditioning. The use of an indirect evaporative cooling system in conjunction with a mechanical air conditioning system offsets cooling loads and significantly reduces energy consumption during peak design conditions.

Hydraulic Scheme

The scheme of an indirect evaporative cooling system is shown in Figure A.14.

Control

The direct evaporative cooler component provides an added cooling benefit only when the ambient dry-bulb temperature is less than the desired supply air temperature. This is because a direct evaporative cooler adds moisture to the supply air as it cools. Similarly, indirect evaporative cooler provides a cooling

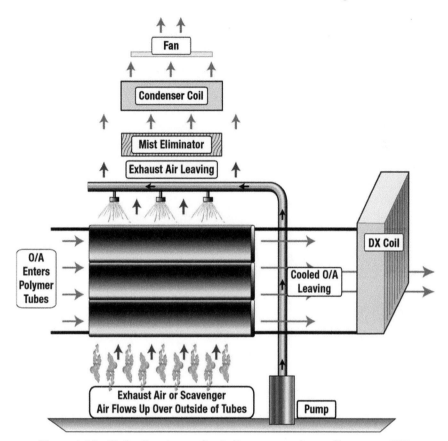

Figure A.14 Hydraulic scheme of an indirect evaporative cooling system [32].

benefit only when the return air wet-bulb temperature is less than the dry-bulb temperature of the outdoor air entering the indirect evaporative cooler. Thus, the system controls typically turn off direct evaporative cooler when the ambient dry-bulb is greater than 13°C, and indirect evaporative cooler is normally disabled when the ambient dry-bulb is less than 17°C to 18°C [44].

Face and bypass dampers on the air-to-air heat exchanger are essential for best control of any two-stage evaporative cooling system, as they allow for economizer cooling and full modulation of indirect evaporative cooler, while similarly allowing for optimal direct evaporative cooler control by regulating the amount of heat recovery as needed to achieve tighter supply air temperature control.

Calculation

The performance of this system is always associated with the other elements of the cooling system. Specific software tools have to be used for selecting the adequate dimension and number of cooling towers for given temperatures and cooling capacity.

Limits of Application

Annual dry- and wet-bulb temperatures at the specific location have to be suitable for the application of evaporative cooling. Additionally, make-up water has to be available.

Economic Aspects

DIEC modules are relatively inexpensive, typically costing 0.37 € per l/s to 0.73 € per l/s for a high-quality unit [36].

A.4.4 Dry Cooler
General Description

A dry cooler (DRC) is a heat exchanger used for transferring heat from a liquid circuit (water or brine) to the ambient air. As shown in Figure A.15, fans are used for blowing the air through the heat exchanger.

Dry coolers can be used for re-cooling the brine circuit of chillers as well as free cooling of chilled water circuits.

Hydraulic Scheme

The hydraulic scheme of a dry cooler plant is shown in Figure A.15.

Control

A set point is given for the brine supply temperature $T_{\text{cow,s}}$. The cooler fans are driven with variable frequency or at least switched between off and two stages (e.g., depending on the ambient temperature). The single coolers with their pumps are switched on and off according to the required cooling power (cascade).

Calculation

The required re-cooling power (heat rejection rate) $\dot{Q}_{\text{Con}}$ is calculated, e.g., for the chiller plant. Design cooling water temperatures can be calculated

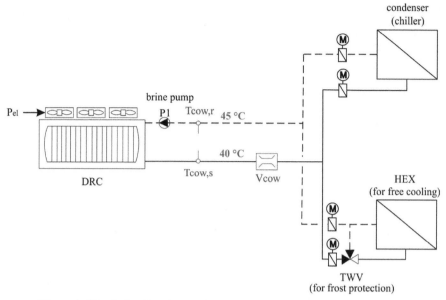

Figure A.15 Hydraulic scheme of a dry cooler plant (example temperatures).

as follows: the design dry-bulb temperature $T_{\text{db,d}}$ depends on the location. Cooling water supply temperature $T_{\text{cow,s}}$ is then known from a given approach (temperature difference between water exiting the cooler and ambient air) which generally is 4–7 K [21]. Fixing the range (temperature difference between water inlet and outlet; typical values are 5–7 K [21]) gives the design water return temperature $T_{\text{cow,r}}$.

Key data of the cooler are the overall heat transfer coefficient U_0 (design value) and the heat exchanger surface area A. The product UA required for the design cooling capacity $\dot{Q}_0$ can be calculated from [50]

$$\dot{Q}_C = \left(1 - \exp\left[\left(\frac{C_{\text{max}}}{C_{\text{min}}}\right)\left(\frac{UA}{C_{\text{min}}}\right)^{0.22}\right.\right.$$
$$\left.\left.\left\{\exp\left[-\frac{C_{\text{min}}}{C_{\text{max}}}\left(\frac{UA}{C_{\text{min}}}\right)^{0.78}\right] - 1\right\}\right]\right) C_{\text{min}} \left[T_{\text{cow,r}} - T_{\text{air,in}}\right]$$

with C_{max} and C_{min} representing the maximum and the minimum value of the capacity flow, respectively. Capacity flow C is calculated as $C = \dot{m}c_p$ from mass flow $\dot{m}$ and specific heat capacity c_p.

Limits of Application

The maximum dry-bulb temperature at the location must not be too high for providing the cooling water temperature required by the process (e.g., vapour-compression chiller).

Economic Aspects

The investment costs for a horizontal dry cooler with fan motor with star connection can be estimated from [20]

$$C[\text{€}] = f_{\text{RPI}}^{(y-2002)} \left(46.169 \cdot \dot{Q}_{\text{rating}} \; [\text{kW}] + 525.02 \right)$$

$$(\text{valid for } 20 \; \text{kW} \leq \dot{Q}_{\text{rating}} \leq 950 \; \text{kW})$$

with $\dot{Q}_{\text{rating}} = \dot{Q}_0 / f_t$ and the correction factor

$$f_t = -0.1 \, t_{\text{air}} \, [^\circ\text{C}] + 3.5 + \left(\frac{t_{\text{brine,in}} \, [^\circ\text{C}] - 40}{10} \right).$$

The power demand of fans and brine pumps can be estimated from the following key figures [20]:

- Fan power: $P_{\text{fan}} = 0.0285 \, \dot{Q}_{\text{rating}}$
- Pump power: $P_{\text{pump}} = 0.0061 \, \dot{Q}_{\text{rating}}$

A.4.5 Wet Cooling Tower

General Description

An open wet cooling tower (WCT) is a heat exchanger used for transferring heat from a water circuit to the ambient air. As the water is in contact with the air, evaporation takes place, which dominates the heat transfer process. As shown in Figure A.17, fans are used for blowing the air through the cooling tower.

A typical wet cooling tower plant consists of several cooling tower cells, which are connected to cooling water basins collecting the warm and the cooled water. Equipping each cell with a cooling water pump is recommended for running arbitrary combinations of cooling tower cells according to demand.

Wet cooling towers can be used for re-cooling the cooling water circuit of chillers as well as free cooling of chilled water circuits.

The enthalpy (*h-x*) chart (Figure A.16) shows the heat transfer processes between air and cooling water in a wet cooling tower.

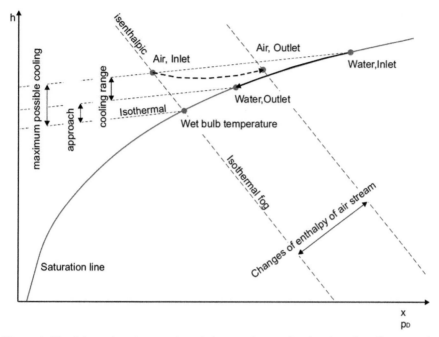

Figure A.16 Schematic representation of change of state of moist air and cooling water in h-x chart for moist air for an open wet cooling tower [53].

In this chart, a cooling limit represents the theoretically possible cooling of water in a wet cooling tower. When the air is dryer (a lower wet-bulb temperature), it can decrease the temperature of the cooling water more through evaporation. Cooling range represents the actual cooling of the water, and the approach defines the distance to cooling limit. These terms assist in evaluating the quality of a cooling tower.

Hydraulic Scheme

The hydraulic scheme of a wet cooling tower plant is shown in Figure A.17.

Control

A set point is given for the cooling water supply temperature. The airflow through the cooling tower cell is controlled by switching between different fan power stages or by using variable-frequency drives for the fans. When the fan of one cell has reached its maximum power and the cooling water supply temperature is still too high, the next cooling tower cell is switched on beginning with the lowest fan power.

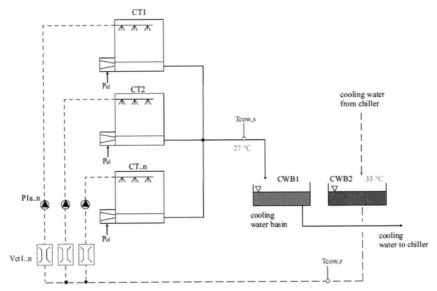

Figure A.17   Hydraulic scheme of a wet cooling tower plant (example temperatures, scheme based on [53]).

An advantage of the basins is their capability to balance fluctuating cooling water temperatures. As a result, a relatively constant cooling water temperature is supplied to the chillers.

Calculation

Obtaining the required cooling capacity and design temperatures is conducted as described in Section A.4.3. As the minimum temperature which can be reached is the wet-bulb temperature, wet cooling tower design is based on a design value of this temperature. It can be, for example, 22°C in Chemnitz, Germany and 26°C in Barcelona, Spain. Typically, the approach and the cooling water range are 4–6 K and 5–7 K, respectively.

Manufacturer software tools have to be used for selecting the adequate cooling tower for given temperatures and cooling capacity.

The power demand P_{fan} of the fans can be estimated depending on the fan type [20]:

$$
P_{\text{fan}} = \begin{cases} 0.0043\, \dot{Q}_{\text{C}}^{1.2336} & \text{for centrifugal fans} \\ 0.0105\, \dot{Q}_{\text{C}}^{0.9613} & \text{for axial fans} \end{cases}
\tag{A.2}
$$

The power demand of the cooling tower pump can be calculated from Equation (A.2) depending on the pressure height Δp.

Limits of Application

Make-up water availability is a requirement for wet cooling tower application. The total water loss sums to about 2.5 to 4.5 l/kWh cooling [27]. Local legislation may restrict the application of wet cooling towers. The effectiveness of re-cooling the cooling water decreases with increased ambient air humidity.

Economic Aspects

The specific investment costs of wet cooling towers can be estimated as [20]

$$c[\text{€}/\text{kW}] = f_{\text{RPI}}^{(y-2002)} \left(2348.2 \, \dot{Q}_C \, [\text{kW}]^{-1.0398} + 26.15 \right)$$
$$\left(\text{for } 50 \ \text{kW} \leq \dot{Q}_C \leq 1.2 \ \text{MW} \right).$$

A.5 Energy Storage

A.5.1 Electrical Energy Storage

A.5.1.1 Electrochemical batteries

Power stability and power bridging are the two roles covered by electrical energy storages commonly available in Data Centres. An increased energy storage capacity together with advanced control algorithms can enhance energy management strategies easing the integration of renewable energy sources in Net Zero Energy Data Centres.

The commonest electrical storage systems are based on batteries even if flywheels and ultra-capacitors have been tested to store energy in Data Centre environments. Accordingly, following sections concern basic descriptions and key aspects for assessing the modelling of the aforementioned electrical storages.

General Description

A battery is an electrochemical device that stores energy and then supplies it as electricity to a load circuit. Batteries are typically organized in strings and can be connected in series, in parallel or in combination of both, in order to provide the required operating voltage and current.

Electrical Scheme

The electrical scheme of an energy storage system can vary according to the Data Centre energy supply architecture and according to the different types of loads which the energy storage system has to supply. If the energy storage is supposed to supply just the IT load, the electrical storage system can be seen as an additional energy capacity that increases the autonomy of the traditional UPS battery. Otherwise, the electrical storage system should be able to supply both normal and UPS loads.

Control

Charging and discharging processes for the battery are controlled by the AC/DC power converters connecting the storage device with the rest of the electrical system of the Data Centre. In particular, one common practice is to use the DC/AC inverter based on controlled transistors (right converter in Figure A.18) to actively control the currents flowing from/to the battery cell to feed the electrical loads of the Data Centre. This controller can be a proportional type ensuring constancy of the electrical frequency within the electrical grid of the Data Centre. In turn, the left-side converter in Figure A.18 can be non-actively controlled, i.e. based on passive diodes. That is because its duty is just to rectify AC voltage from the main grid to be able to connect the direct current terminals of the battery cell. This converter does not control power flows. However, considering bidirectional power flows between Data Centre and the main grid, both converters should have to be controlled, so they are to be based on transistors.

Calculation

The efficiency of the energy storage system is the key parameter characterising its performance. Round-trip efficiency is affected by both charging and

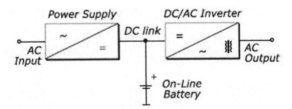

Figure A.18 Battery connected to the DC link of an AC/DC/AC power conditioning unit (topology usually concerned for UPS applications).

discharging operations. In addition, auxiliary loads affect the round-trip efficiency of the system that can be defined as follows:

$$\eta = \frac{E_{\text{out}}}{E_{\text{in}}}$$

where E_{in} is the energy absorbed from an energy source or from the grid while E_{out} is the energy supplied to the grid or to a load. Round-trip efficiency can vary from 95% to 60% according to the cell type.

Limits of Application

Energy storage systems based on batteries are characterised by their great scalability and modularity. This way, designed solutions can be scaled up to several tens of megawatts of power with energy capacity enough to provide energy for several hours at full power. The application of electrical storage in Data Centres is thus not constrained technically, at performances and operating conditions fit with requirements of such installations. Major limits of application are related to economic aspects, as in general terms batteries are expensive components and detailed cost analyses are needed to assess their installation.

Economic Aspects

Table A.4 lists average cost of an electrical energy storage system based on different battery technologies.

Academic work reported by [37] proposes a mathematical expression to estimate the investment cost of electrical energy storage systems. This describes investment cost as a function of power and energy capacities as

$$C[\text{€}] = \text{ce}\left[\frac{\text{€}}{\text{kWh}}\right] \cdot E[\text{kWh}] + \text{cp}\left[\frac{\text{€}}{\text{kW}}\right] \cdot P[\text{kW}],$$

where ce is the cost in terms of energy capacity. It mainly refers to the cost of the "container" of the energy stored, i.e. the electromechanical cell in secondary

Table A.4 Cost parameters for different types of batteries [15, 19]

	Li-Ion	Lead-Acid
Investment cost per unit installed power [€/kW]	200	150
Investment cost per unit energy (cost/capacity/efficiency) [€/kWh]	500	200

batteries. *E* is the energy capacity of the storage system and *cp* is the specific cost in terms of power capacity. It mainly refers to the power conditioner unit for the electrical power exchanged by the storage device at its connection point. *P* is the power capacity of the system.

A.5.1.2 Rotary UPS (flywheels)

General Description

Rotary UPS systems are based on flywheels. Flywheels are electromechanical energy storage systems, which store kinetic energy in a rotating disk being mechanically coupled to an electrical machine spinning at high velocity. When energy is required for the loads, the flywheel is slowed down. Thus, part of the stored kinetic energy is translated into electrical energy through the electrical machine. Conversely, when charging the storage device, the flywheel is accelerated so that the electrical energy consumed by the electrical machine during this acceleration is translated into an increment in the rotating speed of the disk. An electronic power converter drives the electrical machine to control its rotating speed, i.e. to control the state of charge of the energy storage device.

Flywheels are characterised by presenting high cyclability, high ramp power rates, short time responses and high power capacity. On the other hand, the energy capacity of flywheels is very limited and they present high standing losses.

Since the energy capacity of flywheels is relatively low, they are able to exchange their rated power for no more than a few seconds at most until being completely discharged.

Rotary UPS systems mainly concern two major design concepts:

a. The rotary UPS and the generator are separated. In this case, the rotary UPS can have a flywheel as a backup.
b. Combined machine of the diesel generator and the flywheel; DRUPS concept/diesel rotary UPS.

In general, diesel generators are placed as close to the Data Centre as possible, mainly to limit the cable length (limits costs, resilience and vulnerability). It should be taken care of that the generator does not cause any vibrations that influence the Data Centre negatively. The generator should be placed inside a compartment that is fire resistant for 60 min. Finally, it is worth noting that the generator is preferably coupled to the Data Centre on the low voltage level.

Electrical Scheme

According to [37], while applied in UPS systems, a flywheel usually provides the energy required for the transition between the mains failure and the starting of long-term backup systems such as diesel generators. This will permit to maintain electrical frequency within admissible limits ensuring proper security of supply for the loads. As previously noted, diesel generators can optionally be part of flywheel-based UPS systems (DRUPS concept).

For short disturbances (up to a few seconds), the flywheel provides the required energy to the loads. Since most of the mains failures last for no more than a few seconds, diesel systems are activated very few times during the life span of the system, prolonging the life of electromagnetic clutches and other critical components. The electrical scheme is shown as follows. Optional components (electrical motor generator, the clutch and the backup diesel generator) are shown in grey.

As shown, in the simplest configuration, the flywheel is composed by the rotating disk coupled to a permanent magnet synchronous machine and an AC/AC power-conditioning unit. This can be directly connected to the power distribution system of the Data Centre or to a motor generator that in turn can be activated by a backup diesel generator, when the energy stored in the flywheel is exhausted. The aim of the electrical motor generator, apart from translating the mechanical energy of the axis of the diesel generator into electrical power, is to protect the power electronics of the flywheel system against short-circuit currents and to improve the quality of the power exchanged with the rest of the system filtering harmonics and correcting power factor, among other aspects.

Control

UPS system is always connected for safety and regulatory issues. Only non-critical loads are directly connected to the main grid. For those loads fed by UPS, and in normal operating conditions, electrical power flows from the main grid, the safety breaker and through the passive filter for power quality issues, as indicated in Figure A.19. Thus, no power flows from/to the flywheel but it is always in charged state, so spinning steadily. Nevertheless, active filtering of power supply can also be performed by controlling fast power exchange by the flywheel at its point of connection.

In case of a mains failure, the safety breaker (see Figure A.19) is rapidly opened and the flywheel feeds the loads. In this case, flywheel is commanded to be discharged, so it slowed down for several seconds until the connection of backup diesel generators [37].

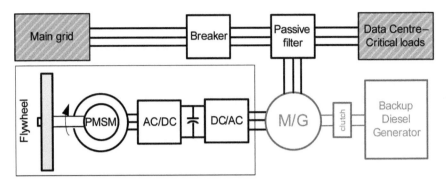

Figure A.19 Flywheel-based UPS system (simplest configuration in black, addition of a backup diesel generator is optional).

Regarding the control of the electronic power converters of the flywheel, it is worth noting that the left-side AC/DC converter in Figure A.19, commonly called machine side converter, is in charge of regulating the speed of the flywheel, i.e. its state of charge. Therefore, it is in charge of controlling the power exchanged by the flywheel. The right-side DC/AC converter is in charge of simply maintaining a constant voltage in the DC link of this set of converters. Thus, both converters have to be based on controlled transistors.

Calculation

For energetic balances, the efficiency of the energy storage system is a key parameter. In this sense, the term "round-trip efficiency" considers both charging and discharging processes and is calculated as

$$\eta = \frac{E_{\text{out}}}{E_{\text{in}}}$$

where E_{in} is the energy absorbed from an energy source or from the grid while E_{out} is the energy supplied to the grid or to a load. For flywheel-based UPS systems, round-trip energy efficiency reaches up to 95%. The simplest configuration for flywheel-based UPS is considered here and corresponds to what is drawn in black in Figure A.19.

Limits of Application

Energy storage systems based on flywheels are characterised by their great scalability and modularity. In addition, the high specific power of flywheels reduces footprint; so designed solutions can reach several MW of rated power even in limited spaces.

Concerning their application in Data Centres, flywheel-based UPSs should be water-cooled or air-cooled, even connected to the building's chiller water supply. Although cooling requirements are marginal for high-tech units, this is something to take into account.

However, as in the case of electrochemical batteries, major limits of application are related to economic aspects, due to the relatively high cost of the system.

Economic Aspects

Table A.5 presents cost estimations for flywheel systems. The simplest configuration (Figure A.19) for flywheel-based UPS is considered here.

According to an industrial benchmark, a 100 kW UPS based on flywheel system (simple configuration, without diesel generator) with energy capacity of about 300 Wh can cost around 60,000 € . A flywheel system equipped with a backup diesel generator with rated power of about 400 kW can cost around 200,000 € .

A.5.1.3 Static UPS

General Description

The static UPS is called "static" because, throughout its power path, it has no moving parts. The UPS raises the level of safety of the IT equipment. The static UPS has three major subsystems: rectifier, batteries and inverter.

UPSs vary greatly in physical size, weight, capacity, supported input power source, technological design and cost. The four main systems are [38] as follows:

- Standby systems – used for power range of 0–5 kVA
- Line interactive UPS – used for power range of 0–5 kVA
- Online UPS double conversion – used for power range of 5–5000 kVA
- Online UPS delta conversion – used for power range of 5–5000 kVA

Table A.5 Cost parameters for flywheel systems [12]

	Flywheels
Investment cost per unit installed power [€/kW]	600
Investment cost per unit energy (cost/capacity/efficiency) [€/kWh]	1200

Here, only the two last-mentioned systems are described in order to compare 3-phase UPSs that can support Data Centres.

In double conversion UPS (Figure A.20), the critical load is provided with fully conditioned mains power, converting AC to DC and back into clean AC energy. The rectifier and inverter constantly convert all the power required by the load. When a failure of the input AC occurs, this does not cause an activation of the transfer switch, because the input AC is charging the backup battery source that provides power to the output inverter. Thus, during an input AC failure, online operation results in no transfer time and the output voltage is of high quality.

The delta conversion UPS (Figure A.21), under conditions of AC failure or disturbances, exhibits behaviour identical to the double conversion, and it was introduced to eliminate its drawbacks. The inverter powers the load constantly, but the delta converter helps to provide power, drawing it from the main power. In this way, the input voltage is treated and integrated with what is missing at the desired value. The delta converter controls the input power characteristics and controls input current in order to regulate charging of the battery system. Delta conversion provides dynamically controlled a power factor corrected input without the inefficient use of filter banks associated with traditional solutions. The most important benefit is a significant reduction in energy losses.

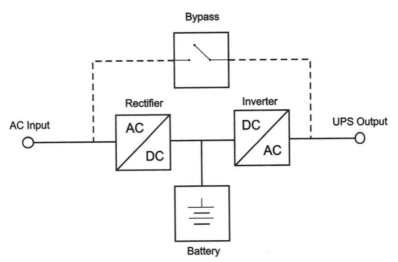

Figure A.20 Double conversion UPS scheme: rectifier and inverter constantly convert all the power required by the load; bypass line is used in case of UPS failure or maintenance.

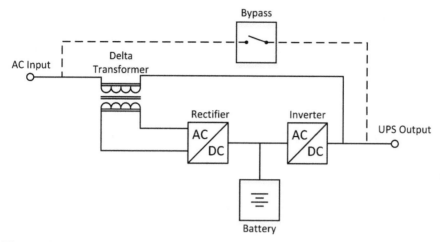

Figure A.21 Delta conversion UPS scheme: rectifier and converter constantly convert all the power required by the load; delta transformer helps to provide power, drawing it from the main power; bypass line is used in case of UPS failure or maintenance.

These systems have several possible configurations. The five most popular UPS configurations are [30] as follows:

- (*N*) System: The UPSs are one or more in parallel, but the overall capacity is never greater than the load to serve. UPSs can be different.
- Isolated redundant: It has two UPSs: one primary that powers the load alone and a secondary that powers the bypass of the primary. This configuration is redundant, when the primary's bypass is activated, and the secondary should be able to support the entire load instantaneously. UPSs can be different.
- Parallel redundant or (*N* + 1) system: More identical UPSs modules in parallel on a single-bus output. It needs a logic control, because it must have synchronised outputs. For this reason, UPSs have to be identical. The load is uniformly distributed among the modules, and if they work at high percentages of load, the total efficiency increases. Every module can be hot swappable. The bus can be a single point of failure.
- Distributed redundant: Three or more UPS modules with independent input and output power lines. If using static transfer switch (STS), the outputs of the UPS must be synchronised. The STS has two inputs and one output to the PDU. UPSs need a controller, because there is a risk during battery operation that the outputs are out of phase. Every module can be hot swappable. STS is a single point of failure.

- System plus system: This double parallel redundant system $(2 \cdot (N + 1))$ eliminates all single points of failure and maintains redundancy even when maintenance is performed. The two power supply lines must follow separate paths. The parallel UPS must be placed in distant locations.

Each configuration provides different levels of availability, performance and costs.

Electrical Scheme

The electrical schemes of the double and delta conversion UPS are shown in Figures A.20 and A.21, respectively.

$$\eta = \frac{E_{out}}{E_{in}}$$

Calculation

Evaluation parameters to size a UPS are operating parameters, availability and performance (efficiency).

The operating parameters are as follows:

- Nominal apparent power S must always be equal or higher than that total load.
- Nominal active power P must always be equal or higher than that total load.
- Overload, necessary to quantify and verify that the UPS can support it.
- Operating temperature tries to predict the temperature rise with the UPS at full load.

The availability defines the ability of the UPS to provide power to the loads continuously and is defined by the following formula:

$$A = (1 - MTTR/MTBF)*100$$

where MTTR is mean time to repair and MTBF is mean time between failures [10].

The efficiency η defines the relation between the output and the input active power at the UPS (Figure A.22).

$$\eta = \frac{P_{out}}{P_{in}}$$

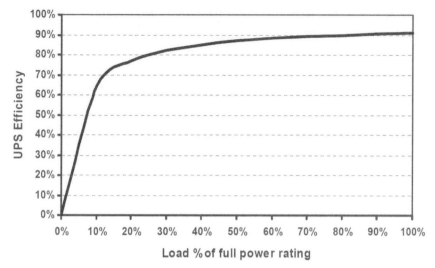

Figure A.22 UPS efficiency decreases with decreasing load, thus better design and install modular UPS.

Limits of Application

There are some limits to be considered, for example, the noisiness. The noise level produced by the UPS should not exceed the allowable average noise level. The inclusion of the UPS in a specific environment must not alter the living conditions of that area.

The size of a UPS is also an important factor. The space required for the installation of UPS is an important parameter for maintainability, and it is a function of space cost per m^2.

The battery life depends on the ambient temperature, the charging parameters and the number of cycles of charge and discharge.

Economic Aspects

A basic utility cost model of a static UPS is described below (Table A.6):

$$\text{Cost of Energy} = P \cdot \left(\frac{1}{\eta} - 1\right) \cdot t \cdot c$$

where P is the active power (kW) supplied to the loads, η is the UPS efficiency reported to a given load level and, therefore, not necessarily the rated efficiency of the machine, t is the time in hours per year of service, to the same load level, and c is the unit cost of electricity per kWh.

Table A.6 Cost parameters for batteries systems [12]

	Batteries
Investment cost per unit installed power [€/kW]	400
Investment cost per unit energy (cost/capacity/efficiency) [€/kWh]	200

A.5.2 Thermal Energy Storage

It is well known that thermal energy storage (TES) could be the most appropriate way to close the gap between thermal energy demand and supply. TES can be based on four principles: temperature difference (sensible TES), phase change (latent TES), sorptive processes and reversible chemical reactions. The energy density in sensible heat storages is determined by the specific heat capacity of the storage material and the temperature difference, while in the latent storage, it is determined by the latent heat (enthalpy of fusion). Due to the still low degree of maturity of sorptive and chemical storages, this study only focuses on using sensible and latent heat storage materials.

There are two kinds of storage: short- and long-term storage. Hence, two different storage strategies can be studied:

Short-Term Storage

The cold is stored for some hours (i.e. during the night and the early morning free air cooling is used to cool the Data Centre, while the TES system is storing cold at the same time). When the outside air conditions do not allow operating in free cooling mode, the cold stored is used to reduce partially or totally the Data Centre cooling demand (Figure A.23a).

Moreover, another strategy is to run chillers when their *COP* is high (the lower the outside temperature, the higher the *COP*) or when the electricity cost is low (during peak-off tariff rates) to produce cold and charge the TES. Notice that from using this strategy, not only energy savings can be accomplished (higher *COP* rates), but also lower operational costs since cheap electricity is used.

Seasonal (Long-Term) Storage

The cold is stored for many weeks or months during the cold season, and it is released into Data Centres during warm/hot periods (Figure A.23b). Although long-term storage has greater potential in practical applications, it is more technologically challenging than short-term storage since it requires large storage volumes and has greater risks of heat losses. Moreover, the storage material must be economical, reliable and ecological.

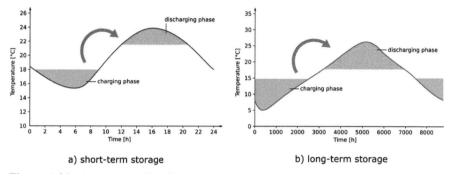

a) short-term storage   b) long-term storage

Figure A.23   Average outside air temperature and TES charging/discharging periods in Barcelona (Spain): spring day with short-time storage (a) and whole year with long-term storage (b).

A.5.2.1 Thermal Energy Storage – Water

General Description

A thermal energy storage (TES) can be applied for short-term storing of chilled or hot water, i.e. as cold or heat storage, respectively. Thermal stratification is important for efficient storage operation. The temperature-depending density of water is used to maintain layers with different temperatures within the storage tank (water temperature decreases from top to bottom). There are different possibilities for storage construction [53]. Here, flat-bottom tanks are assumed for the storage of chilled water.

The storage system consists of the storage tank itself as well as pumps and valves for charging and discharging the storage (see Figure A.24). Connecting the storage to a hydraulic separator between the supply system (chillers for the case of cooling) and the consumer circuit (e.g., cooling distribution system) is favourable.

Hydraulic Scheme

The hydraulic scheme of a chilled water thermal energy storage is shown in Figure A.24.

Control

The parallel pumps (cascade) which are equipped with variable-frequency drives control charging or discharging power. Valves are used for shifting between charging and discharging (these valves are not shown in Figure A.24).

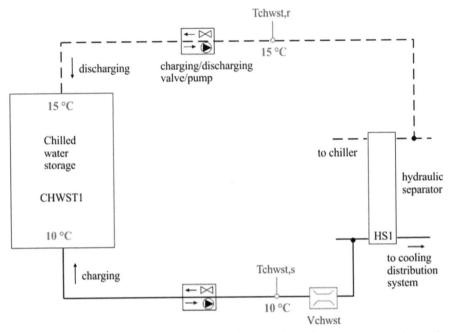

Figure A.24 Hydraulic scheme of a chilled water thermal energy storage (example temperatures) based on [53].

Calculation

For sizing the storage, the required discharging or charging power $\dot{Q}_{CDS}$ has to be defined as well as the time t_{CDS} which has to be covered until the storage is fully charged or discharged. From $\dot{Q}_{CDS}$, the design charging/discharging volume flow $\dot{V}_{CDS}$ can be calculated as

$$\dot{V}_{CDS} = \frac{\dot{Q}_{CDS}}{\Delta T \, \rho \, c_p}$$

with the temperature difference ΔT, water density ρ and specific heat capacity c_p. The required storage volume is obtained from $V_{TES} = \dot{V}_{CDS} \, t_{CDS}$.

The power demand of the charging/discharging pump can be calculated from Equation (A.1) on page 13. The pressure difference Δp depends, for example, on the static pressure in the cooling supply system. For a typical chilled water storage ($V_{TES} = 2700$ m^3, $\dot{V}_{CDS} = 315$ m^3/h), the required pressure difference is $\Delta p_{pump} = 0.\,4$ bar (static pressure 1.5 bar in hydraulic separator).

Limits of Application

Chilled water storages are limited to a minimum water temperature of 4°C due to density rise below this value. Unpressurised hot water storages as considered here are limited to a maximum temperature of 95°C.

Economic Aspects

The specific investment costs of chilled water storage systems ($100 \text{ m}^3 \leq V_{\text{TES}} \leq 6000 \text{ m}^3$) can be estimated as [52].

$$c[\text{€}/\text{m}^3] = f_{\text{RPI}}^{(y-2008)} \left(-119.74 \ln \left(V_{\text{TES}} \left[\text{m}^3\right]\right) + 1291.1\right).$$

A.6 Waste Heat Recovery

A.6.1 (Low-Ex) Heating System

General Description

A building heating system can make use of the heat produced in the Data Centre for space heating, e.g., in offices related to the Data Centre or in other facilities close to it. The heating system is a heat sink characterised by a heat demand and certain supply and return temperatures. Favourable are Low-Ex systems with supply temperatures being close to the ambient temperature.

As shown in Figure A.25, different solutions are possible depending on the IT cooling system:

a. If heat is absorbed from the IT hardware by direct liquid cooling system, the water of the cooling circuit can be used for space heating (heat sink HSI1). A heat storage could be included in the heating system for balancing the difference between heat supply and demand as well as smoothing fluctuating water temperatures.

b. For the case of air-cooled IT, the heated air can be passed directly to the rooms that require heating.

Hydraulic Scheme

The hydraulic schemes of heating systems using heat from the Data Centre for the cases of IT direct cooling and air-cooling are shown in Figure A.25.

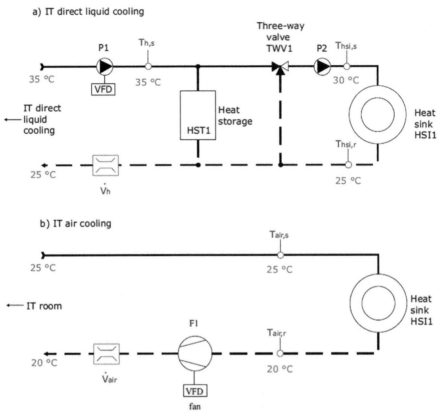

Figure A.25 Hydraulic scheme of heating systems using heat from the Data Centre for the case of IT direct cooling (a) and air cooling (b); example temperatures.

Control

a. Water-heating system

The heating circuit is run with variable flow (pump P2) for energy-efficient operation. The pump control is adapted to the heating system. A three-way valve TWV1 is controlled for maintaining the set point supply temperature $T_{hsi,s}$ (depending on ambient temperature) by admixing return water.

b. Air-heating system

The air volume flow supplied to the heat consumer can be controlled by means of a variable-frequency driven fan.

Calculation

Heat sink inlet and outlet temperatures $T_{\text{hsi,s}}$ and $T_{\text{hsi,r}}$, respectively, are connected to the heating power $\dot{Q}_h$ by

$$\dot{Q}_h = \dot{V}\rho c_p\,(T_{h,s} - T_{h,r})$$

with the heat carrier's flow rate $\dot{V}$, its density ρ and its specific heat capacity c_p. Design temperatures can be for example

- $T_{\text{hsi,s}} = 70°\text{C}/T_{\text{hsi,r}} = 55°\text{C}$ (low-temperature radiator heating system)
- 55/45°C (radiator heating system with reduced temperatures)
- 35/30°C (Low-Ex underfloor heating system)
- 28/22°C (wall and/or underfloor heating system)

Limits of Application

A significant heat demand is required close to the Data Centre. The design temperatures of the IT cooling system and the space heating system have to be harmonised.

A.6.2 Heat Pump

General Description

The main function of a heat pump is raising heat to a higher temperature level in order to supply a heat demand. The main components of an electrically driven heat pump are compressor, condenser, expansion valve and evaporator as shown in Figure A.26. The heat pump cycle is identical to the vapour-compression refrigeration cycle. The basic difference between a heat pump and a refrigerator (chiller) is that in a refrigeration system, cooling of the fluid flowing through the evaporator (primary circuit) is the main purpose, while heating of the fluid in the condenser (secondary circuit) is aspired in heat pump application.

Here, both cooling of the primary circuit and heating of the secondary circuit are beneficial. A water–water high-temperature heat pump is used for cooling the water coming from the white space. The condenser heats water up to, e.g., 70°C for application in space heating systems or feed into a district heating system.

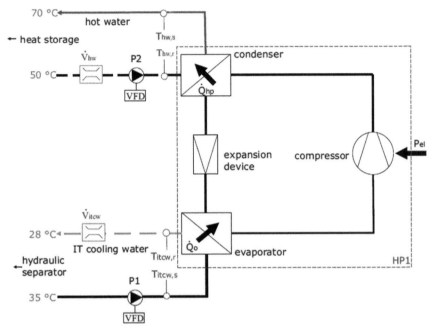

Figure A.26 Hydraulic scheme of a heat pump for water cooling (example temperatures).

Hydraulic Scheme

The hydraulic scheme of a heat pump is shown in Figure A.26.

Control

A set point is presumed for the hot water outlet temperature of the condenser. A pump equipped with variable-frequency drive (VFD) regulates the flow of water flowing through evaporator and condenser accordingly. Controlling of the temperature, load and pressure difference is possible within the allowable limits of the heat pump. Also, desired temperature difference between inlet and outlet of evaporator and condenser could be obtained with the VFD pump.

As cooling supply is very important in Data Centres, controlling the heating power of the heat pump must not result in too low cooling power in the primary circuit.

The coefficient of performance (*COP*) of a heat pump is defined as

$$COP_{hp} = \frac{\dot{Q}_{hp}}{P_{el}},$$

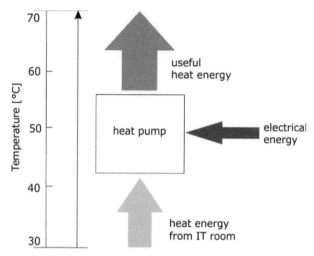

Figure A.27 Schematic energy flow of a heat pump.

where $\dot{Q}_{hp}$ is the thermal output and P_{el} is the power consumption of the unit. In most applications, the *COP* is between 2 and 6 [31]. The *COP* value depends, for instance, on the temperature difference between heat source and heat sink as well as on the type of heat pump. Its value decreases as the temperature difference increases [26].

Figure A.27 shows the energy flow of a heat pump. The thermal output $\dot{Q}_{hp}$ equals the sum of thermal input $\dot{Q}_0$ (cooling power) and electrical energy supply $P_{el} : \dot{Q}_{hp} = \dot{Q}_0 + P_{el}$.

Limits of Application

Water-to-water heat pumps are available in the market from 1 kW to 10 MW [3]. Thermea Energiesysteme [49] offers high-temperature heat pumps with CO_2 as refrigerant in the range from 45 kW to 360 kW (reciprocating compressor) as well as 1 MW (screw compressor). These heat pumps have allowable temperatures at the inlet and outlet of the evaporator from 8 to 50°C and from 2 to 35°C, respectively, and at the condenser from 10 to 50°C and from 40 to 90°C, respectively. The working fluid is water.

Economics Aspects

From budget prices provided by Thermea, the following function was derived describing the investment costs for a high-temperature heat pump (CO_2):

$$c_{\dot{Q}}[\text{€/kW}] = 98370.2 \ \dot{Q}_{hp}[kW]^{-1.2183} + 178.06$$

$$\text{valid for } 45 \text{ kW} \leq \dot{Q}_{hp} \leq 1000 \text{ kW}).$$

Here, $c_{\dot{Q}}$ and $\dot{Q}_{hp}$ represent the specific investment costs and the heating power, respectively. The cost data used for deriving the equation were based on heat source and heat sink temperatures of 35/28°C and 70/40°C, respectively.

A.6.3 Feed into District Heating System

General Description

Heat produced in a Data Centre can be fed into a district heating (DH) system when a DH network is available and suitable temperatures can be reached from direct liquid cooling (with heat pump if necessary). The main component is a heat exchanger (HEX1 in Figure A.28) which separates the Data Centre cooling system from the DH system.

Hydraulic Scheme

The hydraulic scheme of a transfer system for feeding heat into a district heating system is shown in Figure A.28.

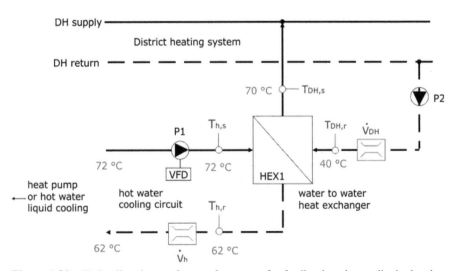

Figure A.28 Hydraulic scheme of a transfer system for feeding heat into a district heating system (example temperatures).

Control

A variable-frequency-driven pump P1 in the hot water "cooling" circuit can be used for keeping the hot water supply temperature at a given set point. However, running the "cooling" circuit according to the supply temperature at the DH heat exchanger must not interfere with IT operation.

Calculation

For a plate heat exchanger operating under counter-flow conditions, the required product of total heat transfer coefficient U_0 and heat transfer surface area A_{HEX} can be calculated from the heat transfer rate $\dot{Q}_{HEX}$ and the mean temperature difference ΔT_m:

$$U_0 A_{HEX} = \frac{\dot{Q}_{HEX}}{\Delta T_m} \text{ with}$$

$$\Delta T_m = \frac{(T_{hot,in} - T_{cold,out}) - (T_{hot,out} - T_{cold,in})}{\ln\left[(T_{hot,in} - T_{cold,out}) / (T_{hot,out} - T_{cold,in})\right]}$$

With the temperatures indicated in Figure A.28, the equation for ΔT_m reads

$$\Delta T_m = \frac{(T_{h,s} - T_{DH,s}) - (T_{h,r} - T_{DH,r})}{\ln\left[(T_{h,s} - T_{DH,s}) / (T_{h,r} - T_{DH,r})\right]}.$$

A typical approach temperature (temperature difference between hot outlet and cold inlet) of water/water heat exchangers is 5 K. However, here a large heat exchanger with an approach of, e.g., 2 K should be used. Thus, the temperature that has to be produced in the hot water "cooling" circuit by the IT hardware or a heat pump is reduced.

Typical design overall heat transfer coefficients of liquid/liquid plate heat exchangers are $U_0 = 1200 \ldots 4500 \text{ W/m}^2 \cdot \text{K}$ [54]. U_0 decreases when the volume flow rates through the heat exchanger are reduced during part-load condition as can be seen from Figure A.29.

Limits of Application

Feeding heat from the Data Centre into a district heating system requires a network being available or built. The hot water produced in the Data Centre needs to have a suitable temperature.

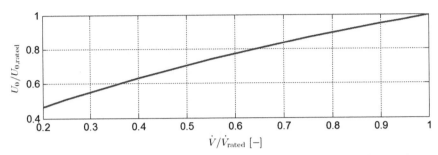

Figure A.29   Relative overall heat transfer coefficient of a water-water plate heat exchanger depending on the relative water flow rate (equal flow rates on primary and secondary side).

Economic Aspects

The investment costs for a gasketed plate heat exchanger can be estimated from [20]

$$C[\text{€}] = f_{\text{RPI}}^{(y-2002)} \left(151.7017\, A_{\text{HEX}}[\text{m}^2]^{0.8829} + 421.17\right)$$
$$\left(\text{valid for } 1 \leq A_{\text{HEX}} \leq 550 \text{ m}^2\right).$$

The cost function was derived for stainless steel plates with EPDM-HT gaskets, maximum temperature and pressure being 160°C and 10 bar, respectively.

Index

About the Editors

Dr. Jaume Salom is the head of the Thermal Energy and Building Performance research group at IREC (Catalonia Institute for Energy Research). Previously to joining IREC in 2010, he co-founded and led the cooperative firm AIGUASOL, which has become an international reference in the field of thermal energy efficiency, renewable energies, building physics and software development. He holds a doctorate degree in Thermal Engineering from the Polytechnical University of Catalonia (Spain) and he has research and professional experience in the fields of heat and mass transfer, fluid mechanics, building energy efficiency, thermal comfort and dynamic simulation. In his current position in the division of Energy efficiency in Systems, Buildings and Communities he leads IREC participation in several national, international and industrial research projects directed towards improving energy performance in buildings and energy systems. Dr. Jaume Salom research interest in the last years has been focused on Net Zero Energy Buildings and Communities, as well of studying energy efficiency and integration of renewable energy sources in Data Centre infrastructures. Dr. Jaume Salom is the Spanish secretary in the TC ISO/IEC JTC1 SC39 *Sustainability for and by information technology.*

Dr.-Ing. habil. Thorsten Urbaneck is the Head of the division of Thermal Energy Storage at the Chemnitz University of Technology since 2006. Since 1996 is scientific staff at this university, hold a doctorate degree since 2004 and has becoming supernumerary professor in 2017. His main research topics are: Thermal energy storages and systems engineering; solar thermal systems; district heating and cooling; combined heat and power; numerical simulation and optimization of system for heating and cooling; evaluation and validation of models in the field of heat and mass transfer; monitoring of supply systems and experiments in laboratory. Thorsten Urbaneck has been publishing more than 190 papers, holding over 170 oral presentations and is the single author of 2 books and co-author of another 2 books. He holds 4 patents and has filed 3 further applications.

Dr. Eduard Oró joined the Catalonia Institute for Energy Research in October 2013, where he currently investigates into the Thermal Energy and Building Performance group. He holds a doctorate degree in Thermal Engineering from the University of Lleida and he has research and professional experience in the fields of heat and mass transfer, fluid mechanics, numerical and dynamic simulation, thermal energy storage systems, transportation and storage of temperature sensitive products. During his stay at the University of Lleida was in charge of the thermal energy storage area at high temperature (>120°C) for solar cooling and concentrated power plants and at the same time at low temperature (<20°C) applications such as transportation of temperature sensitive products such as frozen food, blood, etc. Currently he is coordinating the Green IT area where the implementation of advanced energy concepts and the integration of energy efficiency strategies and renewable energy into data centres are under consideration. In this field, he has participated in some EU funded projects (i.e. RenewIT and Coolemall) supporting the scientific coordination. Dr. Eduard Oró research interest in the last years has been focused on heat reuse potential from data centres as well as studying best integration of these unique facilities in smart cities through smart grids (both electrical and thermal networks).